A Tour of the Subatomic Zoo

A Guide to Particle Physics

❏

Second Edition

A Tour of the Subatomic Zoo

A Guide to Particle Physics

❏

Second Edition

Cindy Schwarz
Vassar College
Poughkeepsie, New York

Introduction by
Sheldon Glashow
Nobel Laureate
Harvard University
Cambridge, Massachusetts

American Institute of Physics   **Woodbury, New York**

In recognition of the importance of preserving what has been written, it is a policy of the American Institute of Physics to have books published in the United States printed on acid-free paper.

AIP Press
American Institute of Physics
500 Sunnyside Boulevard
Woodbury, NY 11797-2999

Library of Congress Cataloging-in-Publication Data
Schwarz, Cindy.
    A tour of the subatomic zoo : a guide to particle physics /
Cindy Schwarz ; introduction by Sheldon Glashow. -- 2nd ed.
    p. cm.
    Includes bibliographical references and index.
    ISBN 1-56396-617-4 (softcover)
    1. Particles (Nuclear physics) I. Title.
QC793.24.S34 1996   96-25726
539.7--dc20   CIP

10 9 8 7 6 5 4 3 2 1

This Second Edition is dedicated
to
My husband Norman
and my two sons, Michael and Bryan

&

CONTENTS

CHAPTER 9: Open Questions   **108**

Appendix   **110**

FIGURES AND TABLES

INTRODUCTION

A Tour of the Subatomic Zoo fills a vast gap in the literature that few of us physicists ever noticed. Physics is a vertical discipline. We do a fine job at training our successors, who are led through a maze of ever more abstract and mathematical disciplines such as mechanics, thermodynamics, electromagnetism, quantum theory, and relativity. Only when they have mastered path integration and dimensional regularization can our students begin to appreciate the newly won understanding of being and becoming at the most basic level. Unfortunately, the 99% of our college students who do not aspire to a career in physics are left entirely out in the cold. They are never taught, and may never learn of, the remarkably simple ingredients of all the wonders of nature. We have irresponsibly denied them an understanding of some of the most profound insights of our time.

Cindy Schwarz provides us with a delightful remedy. In this short, well-worked text (which could form the basis of a six-week course representing a mere one percent of a college education), many of our secrets are revealed to the typical college student and as well to the bright high-school senior or the curious engineer. The book is brief but ambitious. With hardly a mathematical formula, Ms. Schwarz clearly explains the language and much of the substance of elementary particle physics—the search for the ultimate building blocks of matter and rules by which they combine.

Almost a century ago, the electron was discovered: the first of the elementary particles. Since then, the search has brought us today's enormously successful 'standard model' of particle physics. Matter particles (quarks and leptons) interact with one another by various forces, each of which is mediated by the exchange of force particles (photons, gravitons, gluons, W's, and Z's), in accordance with various conservation laws. The preceding sentence should be incomprehensible jargon to intended readers, but will become crystal clear once the book is read and the carefully crafted exercises are completed.

It makes little sense to describe what we know without an explanation of how we know it. The author, herself an active participant in the endeavor, devotes two chapters to a description of the powerful instruments that must be brought to bear to explore the inconceivably tiny seeds of the microworld: giant accelerators and particle detectors. Finally, she tells us some of the many mysteries that remain to be solved by future generations of scientists, and perhaps by Ms. Schwarz herself.

Sheldon Glashow

PREFACE

This book arose from a course on particle physics that I have taught several times at Vassar. A Tour of the Subatomic Zoo is a six-week course intended for students who are not majoring in physics or other sciences and, in fact, assumes no prior physics knowledge. In teaching the course, I found that, although there were many good books on high-energy physics, they were either at too high a level or, if at a proper level, too sweeping. The book serves as an introduction to high-energy physics ideas, terminology, and techniques. Views of matter from the atom to the quark are discussed historically and in low-level technical detail, in a form that an interested person with no physics background can feel comfortable with.

I believe there are several possible audiences for this book. College and university courses could be developed by interested faculty, and this book could be used alone or in conjunction with other materials. Students at Vassar used an early draft, and their feedback was helpful: they found the level to be appropriate and the material interesting. Students often commented that they would not otherwise have been exposed to any physics, as they felt uncomfortable with mathematics. A number of students recommended the course to others. Even college physics majors would enjoy reading this book as an introduction to particle physics. High-school (and even middle-school) teachers could also use the book to introduce some of the material to their students. Many high-school teachers have not been formally exposed to high-energy physics, have forgotten what they once knew, or are no longer up-to-date with recent developments.

Each chapter begins with an overview of the concepts and terms that will be discussed and ends with a summary section. The self-tests at the end of each chapter are not difficult, and answers are included to all the questions. The multiple-choice questions are straightforward and serve as indicators to the reader that basic facts have been learned. The open-ended questions are a little more challenging. Rather than just reading and absorbing a lot of facts, the reader has an opportunity to apply the knowledge he or she has just gained. All the illustrations were done as the text was being written; they are hand-tailored to best illustrate the ideas presented.

Many authors suggest portions of their book that a reader can skip without loss of continuity. One of my goals was to keep things brief so as not to overwhelm or intimidate the reader. My advice, then, is that nothing be omitted. Of course, this means that a number of topics do not appear in this book that many physicists, including myself, feel are important. For example, there is no discussion of the discoveries of radioactivity or quantum mechanics, and many important people are not mentioned. For the reader who wants more, additional books and articles are listed in Suggestions for Further Reading.

This book would never have been written were it not for the students of Vassar College who inspired me to develop the course. I am grateful to them. Thanks also go to Michael Schmidt, Robert Adair, and Jonathan Reichert for carefully reading the preliminary versions. My deepest thanks to my inspiring high-school teacher, Nick Georgis, and my best college professor, Sol Raboy. Without their encouragement and guidance I may never have become a physicist.

Cindy Schwarz
Poughkeepsie, NY

ADDENDUM FOR THE SECOND EDITION

First of all I must thank all of you who purchased and used the first edition of this book. So many of you have told me how much you enjoyed it and how you were using it in your classes (Richard Mancuso and Paul Hickman in particular). This success and events in the field of particle physics over the past five years are my main reasons for revising the book. I have also taught from the book many times and have the Vassar students to thank for their input. In this edition, nothing will be removed, but several things will be added. New developments in high-energy physics, including the March 1995 discovery of the top quark and an updated accelerator table are two of the additions. I have had many conversations (and e-mails) with teachers who want to know about my course and how the book fits in with the whole. For that reason, I am including a copy of my most recent (but always evolving) syllabus in an appendix. I have also included excerpts from some of the papers that the Vassar students enrolled in my course have written over the years. They give an interesting perspective (often what they imagine a particle would say or feel) on the world of sub-atomic physics. Also included will be information on World Wide Web sites where you might find relevant material. Minor typographical and wording errors have been corrected. Thank you to the Modern Physics class of 1993 at the Mississippi School for Mathematics and Science for your helpful comments. I hope that you all enjoy the updated edition. The pictures on Pages 23 and 24 are courtesy of the Contemporary Physics Education Project (CPEP).

Cindy Schwarz
Staatsburg, NY
1996

Chapter 1 MATTER IN THE EARLY 20th CENTURY

In this first chapter we examine views of matter that existed in the early part of the 20th century. When results of experiments could not be explained by known theories, either more experiments were done or new theories were proposed. New theories led to different ideas about what the world around us was made of. Through a look at these experiments and theories, we will learn about some particles that you have probably heard of and some that you may not have heard much about. After completing this chapter you will:

- Know what protons, neutrons, electrons, and neutrinos are.

- Know the constituents of the atom and the nucleus.

- Be able to name three types of radiation.

- Know about conservation rules for energy, electric charge, and momentum.

Parts of the Atom

Rutherford's Experiments: The Nucleus Is Uncovered

We begin with experiments by Ernest Rutherford and his colleagues in Great Britain around 1911. They used a type of radiation, alpha particles, to bombard atoms in an attempt to uncover their inner parts. They did not understand fully what alpha particles were, but they were able to use them. Figure 1-1 shows what their experiment was like. The main points of the experiment were as follows:

- Polonium, which is a radioactive substance, was used as a source of alpha particles. Alpha particles were emitted from the polonium in all directions, but Rutherford was only concerned with the particles that hit the target.

- A movable screen painted with a material called a **scintillator** was used to detect the alpha particles that emerged from the target. Scintillator material gives off a flash of light when struck by an alpha particle. Rutherford was therefore able to study the position of the alphas after they passed through the target.

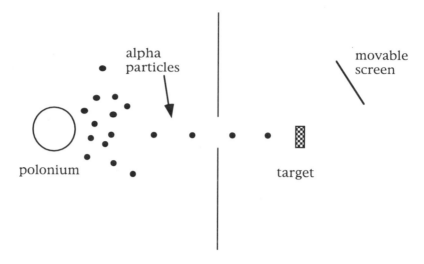

Figure 1-1. Rutherford's experiment.

The results of these experiments were startling. At the time, a popular model of the atom was J. J. Thomson's "plum pudding": a spherically shaped mass of positive charge in which the negatively charged electrons were embedded. (Electrons had been discovered just prior to the beginning of the 20th century by Thomson.) If this model were correct,

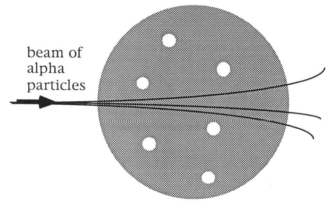

Figure 1-2. Plum pudding.

the results of Rutherford's experiments should have been similar to what is shown in Figure 1-2; the heavy alpha particles would soar through the atom, deflected slightly by their trip through the positive "pudding." To the scientists' surprise, however, some of the alpha particles came right back at them, as if they had bounced off something very massive, and that was inconsistent with the plum pudding model of the atom.

These results, however, were consistent with a new model, seen in Figure 1-3. In this model, the atom contained:

- A solid nucleus in which all the atom's positive electric charge and almost all its mass were concentrated. The alpha particles bounced off this dense nucleus.

- Light electrons existing somewhere in the empty region outside the nucleus. The electrons had a negative electric charge to balance the positive charge of the nucleus. The alpha particles brushed past the electrons.

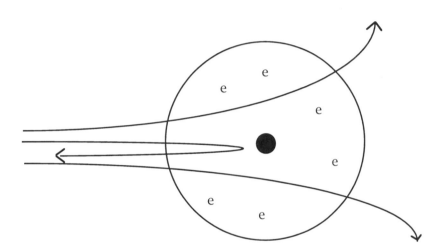

Figure 1-3. Discovery of the nucleus.

But this was not the last of the scientists' experiments. To learn more about the nucleus, Rutherford and James Chadwick continued to use alpha particles. In one series of experiments they shot the alpha particles at nitrogen nuclei and observed the results. As they expected, alpha particles came out, but so did hydrogen nuclei (see Figure 1-4). Well, if hydrogen nuclei could be ejected from nitrogen nuclei, then perhaps nitrogen nuclei were composed of hydrogen nuclei. In fact, perhaps all nuclei were made of hydrogen nuclei.

A **proton** is the name given to a hydrogen nucleus. Protons have one unit of electric charge, equal but opposite to the electron charge. Protons also have mass, and for simplicity we quote all masses in terms of the proton mass (1 unit). In these units, the mass of an electron is about 1/1800.

These experiments showed that nuclei had some sort of internal structure. They too were composed of parts, or protons. The more positive charge a nucleus was found to have, the more protons it must have contained.

Figure 1-4. Alpha particles at nitrogen.

Models of the Nucleus: Protons and Neutrons

The nucleus contains protons. But this is not the complete picture. The simplest model of the nucleus, composed solely of protons and electrons, was suggested in 1914. Let's see how this model works for the nitrogen nucleus. Below is a table with properties of the nitrogen nucleus, the proton, and the electron.

object	charge	mass
proton	+1	1
electron	-1	0
nitrogen nucleus	+7	14

The only way to combine protons and electrons into a nitrogen nucleus is shown in Figure 1-5. Remember, these particles are *in* the nucleus, there will still be seven electrons *outside* the nucleus.

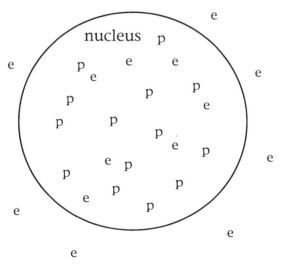

Figure 1-5. Early model of the nitrogen nucleus.

4

Or, in terms of the particles involved, one can see how the charge and mass of the constituents add up to the charge and mass of the nucleus as a whole.

	14 p	+	7 e⁻	=	nitrogen nucleus
charge	14		-7	=	7
mass	14		0	=	14

There is a problem with this model, however, in that the spin is not correct. **Spin** is a property that a particle can have, just as it can have charge and mass. The classical analogy that is most often given for spin is that of a top spinning on its axis. You may think of spin this way, but electrons and protons and all other particles with spin do not actually spin like a top. Spin is an internal property of a particle that can be calculated or measured, just like mass can. Electrons and protons each have a spin of 1/2 unit. The spin of the electron or proton is said to be up or down, and these are the only possibilities. Now, our model has 21 particles in the nucleus, each with a spin of 1/2. The nitrogen nucleus has a spin measured at 1 unit. It is not possible for an odd number of particles with 1/2 unit of spin each (21 in our case) to have a combined spin equal to an integer (1 in our case). This point is illustrated in Figure 1-6 for some simpler cases with three or four particles, each with a spin of 1/2.

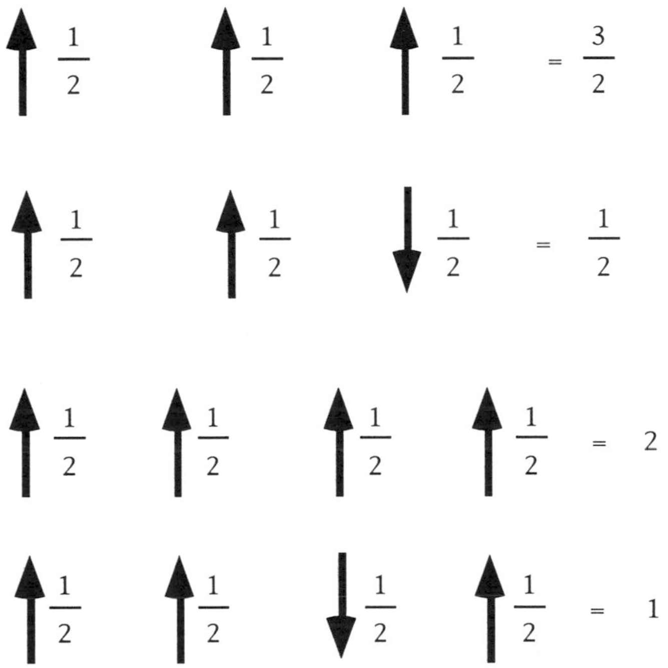

$$\uparrow \tfrac{1}{2} \quad \uparrow \tfrac{1}{2} \quad \uparrow \tfrac{1}{2} \quad = \quad \tfrac{3}{2}$$

$$\uparrow \tfrac{1}{2} \quad \uparrow \tfrac{1}{2} \quad \downarrow \tfrac{1}{2} \quad = \quad \tfrac{1}{2}$$

$$\uparrow \tfrac{1}{2} \quad \uparrow \tfrac{1}{2} \quad \uparrow \tfrac{1}{2} \quad \uparrow \tfrac{1}{2} \quad = \quad 2$$

$$\uparrow \tfrac{1}{2} \quad \uparrow \tfrac{1}{2} \quad \downarrow \tfrac{1}{2} \quad \uparrow \tfrac{1}{2} \quad = \quad 1$$

Figure 1-6. Examples of spin combinations.

By 1930, many physicists realized the inadequacy of the simple model of protons and electrons as the constituents of the nucleus. If we require the nitrogen nucleus to have the correct spin as well as charge and mass, then there must be an even number of objects in the nucleus. So, if we combine 7 of the 14 positively charged protons with the 7 negatively charged electrons in the nucleus, we get 7 neutral objects with masses essentially equal to the proton mass. Figure 1-7 shows how this combination might be viewed. As early as 1920, there were suspicions of a neutral object (with

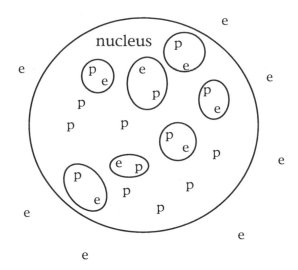

Figure 1-7. Early nuclear model.

the same mass as the proton) in the nucleus. By 1932, the **neutron** had been discovered. Seven protons and seven neutrons in the nitrogen nucleus give the correct charge, mass, and spin. Go ahead and verify it.

The neutron was discovered by Chadwick in another experiment involving alpha particles. A sketch of the basic setup is shown in Figure 1-8. The main points of this experiment were:

- Alpha particles were aimed at a beryllium target.
- Chadwick observed something coming out of the beryllium that did not have electric charge (we use the term **neutral particles** or **neutral radiation**).
- This radiation hit a paraffin target.
- Protons were knocked out of the paraffin target.

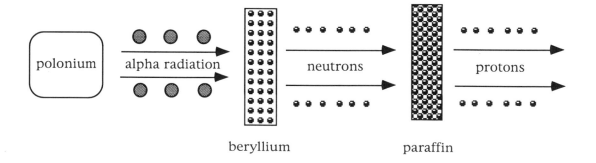

Figure 1-8. Discovery of the neutron.

Chadwick concluded that the neutral radiation was indeed the neutron, as it had to have a mass close to the proton to knock one out of the paraffin.

Radiation

Unstable nuclei spontaneously decay by emitting particles. This process is called **radioactivity**. The three types of radiation are called:

- Alpha (α).
- Beta (β).
- Gamma (γ).

We have already seen that alpha particles were useful in Rutherford's experiments on the atom. Alpha (α) radiation, or alpha particles are simply helium nuclei (two protons and two neutrons). An example of a process that produces alpha particles is a radium nucleus decaying into a radon nucleus and an alpha particle. You can see that the total number of protons and the total number of neutrons remain constant.

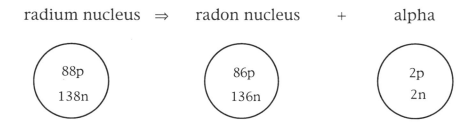

radium nucleus ⇒ radon nucleus + alpha

88p 138n   86p 136n   2p 2n

Gamma(γ) radiation is high-energy electromagnetic radiation. When a nucleus gets into an excited state, it can emit the extra energy it has by gamma radiation. Just to put the energy in perspective, gamma rays prevalent in high-energy physics processes have an energy about 10 billion times that of microwaves. Many different types of nuclei can become excited and emit gamma rays.

excited nucleus ⇒ unexcited (same) nucleus + γ

Beta (β) radiation involves the emission of an electron. But this electron is not one of the electrons from outside the nucleus. It is an electron that is created within the nucleus. An example is:

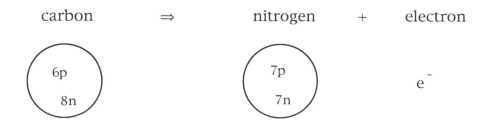

carbon   ⇒   nitrogen   +   electron

6p 8n   7p 7n   e⁻

Unlike the process that produces alpha radiation, in this process the total number of protons and the total number of neutrons are not separately conserved. We have gained a proton and lost a neutron. The underlying process here is a neutron decaying into a proton and an electron. Studying this process of beta decay led physicists to propose a new particle called the **neutrino**. The neutrino was postulated when neutron beta decay did not follow some important conservation principles of physics. The neutrino will be discussed in greater detail after we discuss these conservation principles.

Some Conservation Laws

Conservation principles are important in physics, and they can help us to explain why some things happen and others don't. Three conservation rules that must be obeyed in all physical processes are:

- Conservation of electric charge.
- Conservation of momentum.
- Conservation of total energy.

We will not derive these rules, but we will use them to determine whether certain particle reactions can or cannot occur.

Charge Conservation

The total electric charge of a system of particles must remain the same. That is, if you add up all the charges of the particles on one side of a reaction, they must exactly equal the sum of all the charges of the particles on the other side of the reaction. For example, both the following reactions obey the principle of charge conservation. Note that these are not the only possibilities, but just two examples of reactions.

+	-	⇒	0	0	
positive particle	negative particle		neutral particle	neutral particle	

+	-	⇒	+	-	0
positive particle	negative particle		positive particle	negative particle	neutral particle

In both examples, two charged objects, one with a positive charge and the other with a negative charge, are the initial particles (those to the left of the arrow). In the first case, the final particles (those to the right of the arrow) both have no electric charge. In the second case, there are three particles: one positive, one negative, and one neutral. In either case, the total charge of the final particles is zero.

Energy Conservation

The total energy of the particles before a decay or reaction (two particles in the initial state) must be equal to the total energy of the particles afterward. Now, here we speak of the *total energy*, because the energy that concerns us comes in two forms:

- Kinetic energy, or energy of motion, which depends on the velocity of the particle.

- Mass energy, which arises from Einstein's famous equation

$$E=mc^2$$

In this equation, the energy (E) is equal to the product of the mass of the particle and the speed of light (which is a constant) squared. The greater the mass of the particle, the more mass energy it has. For decays, our criterion for satisfying energy conservation will be that the mass (or mass energy) of the decaying particle be greater than or equal to the sum of the masses of the end products. How does this lead to energy conservation? Energy conservation for particle A initially at rest gives us

mass of A = mass of B + kinetic energy of B + mass of C + kinetic energy of C

and since all energies are positive

(mass of A)   >   (mass of B)   +   (mass of C)

For reactions, the two particles in the initial state can have a total mass less than the total mass of the final particles because they can bring kinetic energy into the reaction to assure that energy is conserved.

Momentum Conservation

The total momentum of the particles must remain the same in any physical process. For velocities that are very low compared to the speed of light, the momentum of a particle is the product of its mass and its velocity. In a particle reaction, if the initial total momentum (the sum of the momenta of the individual particles involved) is zero then the final total momentum must be zero as well. For particle decays, such as one particle becoming two or more particles, this conservation of momentum manifests itself in an interesting visual way. In particular, if one particle decays into exactly two particles then these particles must emerge from the reaction in exactly opposite directions for momentum to be conserved.

This is illustrated in Figure 1-9. In this example

A $\Rightarrow$ B + C

M, the mass of particle B, is larger than m, the mass of particle C. Therefore, the velocity of particle C must be larger than the velocity of particle B (as indicated by the length of the arrows). Equivalently, the momentum of particle B (Mv) is the same as the momentum of particle C (mV).

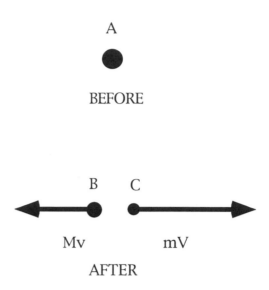

Figure 1-9. Conservation of momentum.

Neutrinos

A reaction that violates one of these conservation principles cannot occur. The process of nuclear beta decay was studied carefully in the 1920s. One of the things that scientists found was that energy in beta decay of the neutron was not conserved. Physicists at this point had a tough choice: They could abandon the principle of energy conservation, or they could accept a new hypothesis put forth by Wolfgang Pauli. Pauli proposed that a new particle, which could not be detected, ran off with the missing energy. This particle was to have no electric charge, little (or no) mass, and the same spin as protons and electrons. It was given the name **neutrino**, which means "little neutral one" in Italian. Existence of this particle was well accepted by physicists by the 1950s.

By the late 1950s, it was also seen that momentum was not conserved in beta decay of the neutron unless the neutrino was considered a part of the process. If the neutron decayed into only two particles, then the diagram on the left in Figure 1-10 shows what one must see to guarantee conservation of momentum. The proton and the neutron must emerge from the decay back-to-back.

10

But that is not what physicists observed in cloud chamber pictures. Instead, the emerging particles looked something like those shown in the diagram on the right in Figure 1-10.

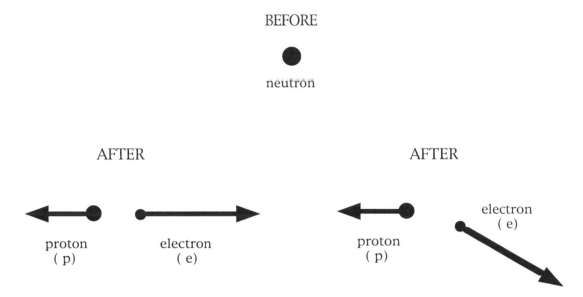

Figure 1-10. Neutron decay possibilities.

The idea of the neutrino "fixed" neutron beta decay so that it obeyed all the conservation rules. However, experimental verification was needed before everyone would believe in the neutrino. In 1956, about 25 years after it was initially proposed, the neutrino was discovered outside a nuclear reactor. At the Savannah River (Ga.) reactor, the number of neutrinos emerging per second was extremely high, and physicists waited with their detector until they eventually detected some.

So the theory of beta decay must be modified to include the neutrino. The symbol for the neutrino is the Greek letter ν, and it has no electric charge and little (or no) mass. Current experiments are placing limits on the mass of the neutrino, but for our purposes it will be taken as zero. This may be the first particle that you have not heard of. To this point we have dealt with the traditional everyday atom, stable or not, composed of protons, neutrons, and electrons. Now we encounter the neutrino, whose existence was postulated when neutron beta decay did not follow important conservation principles. A diagram of the process is shown in Figure 1-11, with the four subatomic particles all linked together in beta decay.

$$n \Rightarrow p + e^- + \nu$$

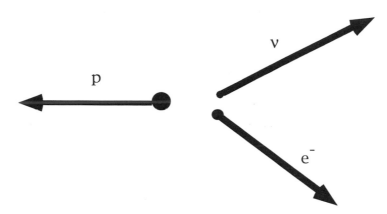

Figure 1-11. Neutron decay.

"Radioactive decay is responsible for my existence today. Some like to say the weak force brought me into this world today, but I prefer to think my mother gave her life for me. Mom was a neutron. After a relatively short life of fifteen minutes, she decayed into her three children. I'm a proton and my two little sisters are an electron and a neutrino. Mom was a model particle. She didn't cause much trouble and stayed out of other particle's business because she was so neutral."
Taegan D. Goddard

12

Summary

In this chapter:

We saw how the experiments done in the early part of the century led to:

- The model of the atom: a dense nucleus containing most of the mass and all of the positive charge, with electrons somewhere outside the nucleus.

- The model of the nucleus: protons and neutrons, with protons having positive electric charge and neutrons having none.

- The discovery of the neutron.

We learned about three types of radiation (α, β, γ) and that the study of one of the types, beta radiation, led to the theory of the neutrino.

We introduced three conservation laws for:

- Charge.
- Momentum.
- Energy.

Self-Test 1

For multiple choice, check all that apply.

1. Rutherford used which kind of radiation to study atomic and nuclear structure?

    a. alpha
    b. beta
    c. gamma

2. Match the items in the left column with those in the right.

    alpha   a neutron changes into a proton
    beta   high-energy electromagnetic radiation
    gamma   a helium nucleus

3. Neutrinos were proposed by Pauli:

    a. because they were seen at a nuclear reactor.
    b. because of missing energy in beta decay.
    c. to guarantee charge conservation in beta decay.

4. Name the three conserved quantities you have learned about so far.

5. Fill in the chart

object	symbol	mass	charge	spin
proton				
neutron				
electron				
neutrino				

6. Test each reaction for <u>charge conservation</u> and conclude whether it passes or fails.

$$e^- + p \Rightarrow n + n$$

$$e^- + p \Rightarrow v + n$$

$$e^- + n \Rightarrow p + n$$

7. Draw a picture of a helium atom as seen by a physicist in 1932.

Answers to Self-Test 1

1. Rutherford used which kind of radiation to study atomic and nuclear structure?

     a. alpha   X
     b. beta
     c. gamma

2. Match the items in the left column with those in the right.

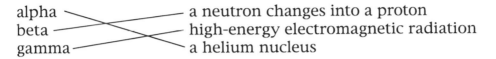

     alpha   a neutron changes into a proton
     beta   high-energy electromagnetic radiation
     gamma   a helium nucleus

3. Neutrinos were proposed by Pauli:

     a. because they were seen at a nuclear reactor.
     b. because of missing energy in beta decay.   X
     c. to guarantee charge conservation in beta decay.

4. Name the three conserved quantities you have learned about so far.

    momentum
    energy
    charge

5. Chart

object	symbol	mass	charge	spin
proton	p	1	+1	1/2
neutron	n	1	0	1/2
electron	e⁻	1/1800	-1	1/2
neutrino	v	0	0	1/2

6. Test each reaction for <u>charge conservation</u>.

$$e^- + p \Rightarrow n + n$$

    -    +   0   0   conserved

$$e^- + p \Rightarrow v + n$$

    -    +   0   0   conserved

$$e^- + n \Rightarrow p + n$$

    -    0   +   0   not conserved

7. A helium atom as seen by a physicist in 1932. Helium has 2 protons, 2 neutrons, and 2 electrons. This and all other drawings in this book are not to scale. The nucleus takes up an extremely small fraction of the total size of an atom. The atom has a diameter about 10,000 to 100,000 times as big as the diameter of the nucleus.

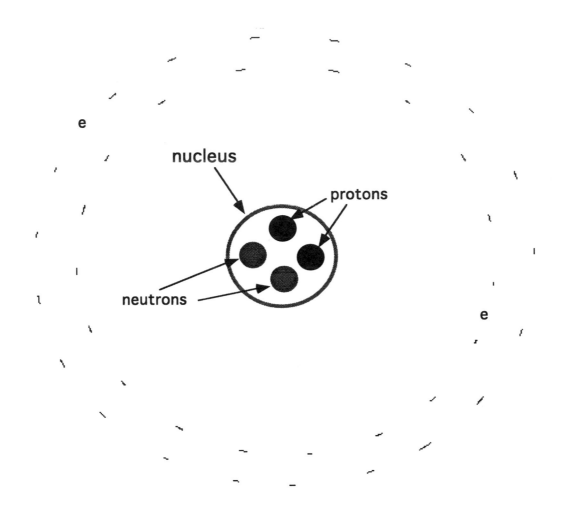

"Oh, this endless spinning. The monotony is overwhelming. I could have been a proton; I could have been a neutrino, but no. I'm am electron; a negatively charged, crazily spinning particle, and I don't even get to choose my path. I'm stuck on this highway and it's one big circle (sort of). I start in the same place, and I end up in the same place, and then I start all over again. Basically, I am leashed onto this nucleus, and I can't get away until some better prospect comes along or until I'm spit out. I just keep going around and around. I've got this serious attraction to my nucleus. There's something so positive about it, so electrically alluring..." Danielle Otis

Chapter 2 FORCES AND INTERACTIONS

A **force** is something that changes the momentum (hence velocity) of an object. Or we can say, equivalently: If an object has a change in velocity (an acceleration), then a force acted on it. In the subatomic world of particles, forces are transmitted by the exchange of intermediary particles, or force carriers. When two particles exert a force on each other, they do it by exchanging a force carrier. A helpful analogy is to picture two people throwing a basketball back and forth. As they throw the basketball (the carrier) they exert a force on each other. In this chapter we will learn about the four forces that are presently considered to be fundamental and the carriers of these forces. They are the:

- Gravitational force.

- Electromagnetic force.

- Strong force.

- Weak force.

We will also learn how the particle processes (called **interactions**) are represented (via these forces) as Feynman diagrams. These diagrams allow physicists to calculate some important quantities in high-energy physics, but we will use them to get a better understanding of the different kinds of reactions that can take place.

Fundamental Forces

Gravitational Force

The gravitational force acts between all particles that have mass. Every mass attracts every other mass with a force that gets weaker as the distance between the masses increases. The gravitational force is the binding force of the solar system and the galaxies, and is therefore a very important force. In the context of high-energy physics, however, it is not important, because the masses of particles are very, very small, and the strength of the gravitational force is proportional to the mass of the objects involved. Nonetheless, all objects with mass do experience the gravitational force even when it is feeble. The carrier of the gravitational force is called the **graviton**, but it has never been observed in an experiment.

Electromagnetic Force

The electromagnetic force acts between all objects that have electric charge. Of the particles that we know about, the electron and the proton have electric charge, and the neutron and the neutrino do not. The force is attractive for oppositely charged objects and repulsive for objects with the same charge. Therefore, two electrons will repel each other, and an electron and a proton will attract each other. It is in this manner that atoms are held together. As with the gravitational force, this force gets weaker as the distance between the particles increases. Since the neutrino and the neutron have no electric charge, they are not affected by the electromagnetic force. The carrier of the electromagnetic force is the **photon** (γ), and it definitely exists. When you see light, you see photons.

Strong Force

The strong force is an attractive force that acts between **nucleons** (the collective name for protons and neutrons). It is attractive for all combinations of protons and neutrons:

- Protons attract protons.
- Neutrons attract neutrons.
- Protons attract neutrons.

Were it not for the strong force, the electromagnetic force of repulsion between the protons in the nucleus would cause the nucleus to break apart. We will see later that the strong force actually acts between quarks, which are the parts of the proton and neutron, but until then we will concern ourselves with the effect of this underlying process on the nucleons. The carrier of the strong force is called a **gluon** (g), and there is experimental evidence for its existence.

Weak Force

The weak force, so-called because it is weak in strength compared to the strong force, is responsible for radioactive beta decay. If you recall, that process involved a neutron turning into a proton, electron, and neutrino. Neutrinos are affected only by the weak force, as they have no mass or electric charge, and the strong force affects only nucleons. So, whenever a neutrino is involved in a reaction, the reaction must proceed under the weak force. The carriers of the weak force are the W^+, W^-, and Z^0. The W's and the Z are very massive. In contrast to the other three force carriers, which have zero mass, they have masses almost 100 times the proton mass.

Interactions and Feynman Diagrams

It is customary to draw the particle interactions as Feynman diagrams. These diagrams are useful to high-energy physicists in that they facilitate complex calculations. These calculations are way beyond the scope of this book, but we can use Feynman diagrams to help us visualize the particle processes. The rules for drawing them are as follows:

- Draw all particles, incoming and outgoing, as straight lines.
- Draw all force carriers as wavy lines connecting the particles.
- Draw arrows to indicate the direction of motion in time.

In Figure 2-1 we see the diagram for two electrons interacting via the electromagnetic force. In this example, the particles that go into the reaction are the same as the particles that come out of the reaction. It is somewhat analogous to a collision between two billiard balls. When two billiard balls collide, they exert equal and opposite forces on each other, which result in changes in velocity. The electrons interact by exchanging a photon and thereby exerting a force on each other. Other, more complex diagrams are possible for the electromagnetic interaction of two electrons, but the simple one-photon exchange (shown below) is the most common process. This is an electromagnetic process because of the presence of the photon as the force carrier. You now have the information to draw some simple diagrams, given the particles involved and the nature of the interaction. But we must learn about more particles before we can go on. Figure 2-2 on the next page shows two more examples.

$$e^- + e^- \Rightarrow e^- + e^-$$

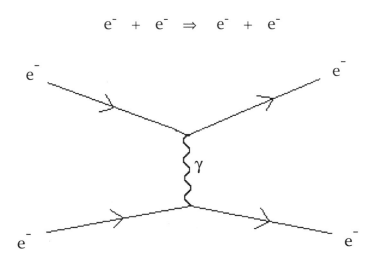

Figure 2-1. Feynman diagram.

ELECTRON-NEUTRINO SCATTERING

$$e^- + \nu \Rightarrow e^- + \nu$$

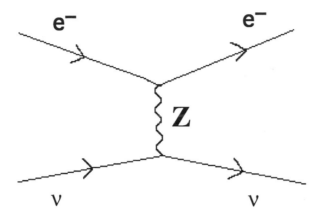

BETA DECAY

$$n + \nu \Rightarrow p + e^-$$

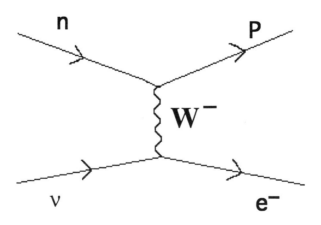

Figure 2-2. More Feynman diagrams.

Summary

In this chapter:

We discussed the concept of force as it applies to subatomic particles. In particular, we learned about the four forces presently thought to be fundamental.

- The gravitational force is attractive for all objects that have mass. The mediator is thought to be the graviton, which has not yet been observed directly or indirectly. It is the binding force of solar systems and galaxies. It is relatively unimportant for subatomic particles.

$$mass = 0$$
$$charge = 0$$
$$spin = 2$$
$$range = infinite$$

- The electromagnetic force acts between electrically charged objects. It can be attractive or repulsive. It is the binding force of atoms. The positively charged nucleus is attracted to the negative electrons. The mediator is the photon, which is known to exist.

$$mass = 0$$
$$charge = 0$$
$$spin = 1$$
$$range = infinite$$

- The strong force is attractive for all nucleon pairs. It is the binding force of the nucleus. The mediator is the gluon, which has been observed indirectly.

mass = 0
charge = 0
spin = 1
range = 10^{-15} m

- The weak force can affect all particles, but it is the only force that affects neutrinos. It is responsible for beta decay. The mediators are the two W's and the Z. They were first observed directly in 1983.

mass = 82 (W), 91 (Z)
charge = +1/-1 (W), 0 (Z)
spin = 1
range = 10^{-19} m

We also learned what a Feynman diagram is and how to draw a simple one. Don't worry, we will get to some more complicated ones in good time.

Self-Test 2

For multiple choice, check all that apply.

1. The gravitational force is

    a. always repulsive.
    b. always attractive.
    c. sometimes attractive and sometimes repulsive.

2. The electromagnetic force

    a. acts on electrically charged particles.
    b. is always attractive.
    c. is always repulsive.

3. The strong force

    a. attracts neutrons to protons.
    b. attracts electrons to protons.
    c. attracts protons to protons.

4. The weak force is

    a. responsible for stability of the nucleus.
    b. responsible for radioactive decay.
    c. the only force that affects neutrinos.
    d. the only force that affects neutrons.

5. Match the force carrier in the left column with the force in the right.

gluon	weak
graviton	strong
W, Z	electromagnetic
photon	gravitational

6. Fill in the missing force carriers.

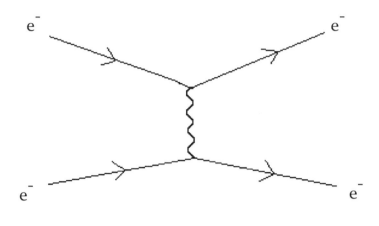

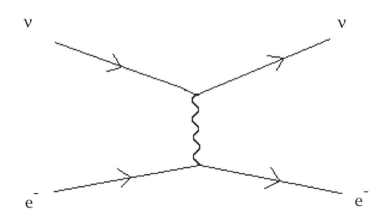

"Who are you, anyway? And why are you tugging on me?" "I'm sorry. My name is Wally and I'm a Weak Force mediator. Please don't hit me! I have to do this, it's my job. I didn't ask to be made part of the Weak Force, you know. My official title is W-, most everyone calls me Wally. Wilma is the W+ and Zachary is the neutral Z Now I have to break up your constituent parts, so don't get mad or throw photons at me!" Jeffrey Gordon

Answers to Self-Test 2

1. The gravitational force is

    a. always repulsive.
    b. always attractive.   X
    c. sometimes attractive and sometimes repulsive.

2. The electromagnetic force

    a. acts on electrically charged particles.   X
    b. is always attractive.
    c. is always repulsive.

3. The strong force

    a. attracts neutrons to protons.   X
    b. attracts electrons to protons.
    c. attracts protons to protons.   X

4. The weak force is

    a. responsible for stability of the nucleus.
    b. responsible for radioactive decay.   X
    c. the only force that affects neutrinos.   X
    d. the only force that affects neutrons.

5. Match the force carrier in the left column with the force in the right.

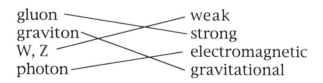

gluon — strong
graviton — gravitational
W, Z — weak
photon — electromagnetic

6. Fill in the missing force carriers.

When the electromagnetic force can mediate an interaction, it will, although it is possible for the Z to be the mediator.

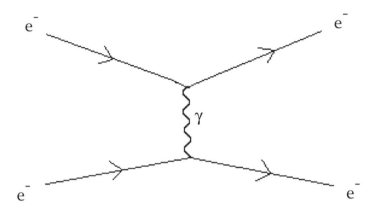

In this case, since neutrinos are involved, the force responsible for the interaction must be the weak force. The possible mediators therefore are the Z^o, the W^+, or the W^-. Since the electron remains an electron and the neutrino remains a neutrino, no charge is exchanged between the particles. The mediator therefore must have no electric charge and is therefore the Z^o.

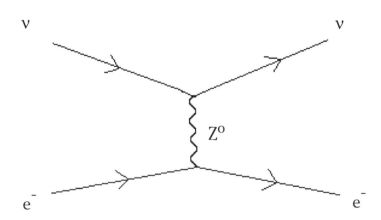

Chapter 3 A GLIMPSE AT THE PARTICLE ZOO

In this chapter, we will discuss some of the experiments that were done in the 1930s, '40s, and '50s. Discoveries from these experiments led to theories that in turn led to more experiments, more discovery, and more experiments. The scientists of the times knew that the atom was composed of electrons and a nucleus, which in turn contained protons and neutrons. But there was no reason to believe that that was the final model. On completion of this chapter, you will be introduced to several new particles, including:

- positrons
- antiprotons
- antineutrons
- pions
- muons
- more neutrinos
- antineutrinos

Antimatter

The Positron

In seeking to combine Einstein's theory of special relativity and the equations governing the behavior of electrons in electric and magnetic fields, Paul Dirac was led to a remarkable idea. His equations had some peculiarities, one of which prompted him to predict the existence of a new particle similar to the electron. This new particle was called the **positron** (e+), and it was to have the same mass and spin as an electron but opposite electric charge. Let me again stress that the existence of this particle was postulated from a theory and was subsequently discovered in an experiment. The positron was discovered in 1932, the same year the neutron was discovered. Because of the symmetry between this particle and the electron it is now also known as the **antielectron**.

Other Antiparticles

Now, there was no reason that electrons should have the distinction of being the only particle with an antiparticle, so antiparticles were proposed for the proton and neutron as well. Antiprotons have the same mass and spin of a proton with opposite (negative) electric charge. Antineutrons have the same mass and spin of a neutron with opposite (zero) electric charge. Something may not seem right with this idea. You can distinguish the proton from the antiproton by its electric charge, but how do you tell the difference between the neutron and the antineutron? The complete answer to this question will have to wait until Chapter 5, but I will tell you now that, although they seem indistinguishable from the outside, the neutron and antineutron are indeed different particles.

The antiproton and the antineutron were not discovered until 1955. Higher-energy particle collisions were needed to produce them because they are more massive than the electron. They were not given new names and the symbols are just bars on top of the symbols for proton and neutron ($\overline{p}$ and $\overline{n}$).

Properties of Antimatter

There are two important properties of matter and antimatter.

- When a property *has* an opposite value, the antiparticle will have the *opposite* value of the property. An example is:

  electric charge (+ and -)

- When a property has *no opposite*, the particle and antiparticle will have the *same* value for the property. Some examples are:

  mass
  spin
  neutral charge

There is one more very important property of matter and antimatter. When matter and antimatter meet in a particle reaction, they can annihilate into energy, so long as it is an exact matter/antimatter pair. Also, energy can become a matter/antimatter pair of particles so long as the energy is greater than the masses of the particles. This energy manifests itself in the form of gamma radiation (or, equivalently, the term **photons** is used). The following processes are examples that can occur:

- A proton and antiproton annihilation.

$$p + \overline{p} \quad \Rightarrow \quad \text{energy} \, (\gamma)$$

- Energy equivalent to at least twice the electron mass produces an electron/positron pair.

$$\text{energy} \, (\gamma) \quad \Rightarrow \quad e^+ + e^-$$

- An electron and a positron can form energy, which in turn can form a proton and antiproton. In this case, though, it is essential that the electron and positron bring enough kinetic energy into the collision to create the proton and antiproton masses.

$$e^+ + e^- \quad \Rightarrow \quad p + \overline{p}$$

But, remember, it must be an exact matter/antimatter pair. The following can never occur.

$$e^+ \;+\; \bar{p} \;\Rightarrow\; e^- \;+\; p$$

(The combination of the positron and the antiproton is very interesting, though. It is anti-hydrogen, true atomic antimatter. As of the writing of this book in 1996, an experiment at CERN has been successful in synthesizing nine atoms of anti-hydrogen. Further experiments will try to make more and study its properties.)

New Particles

Pions

There is a mathematical relationship between the range of a force, that is the distance over which the force is "felt," and the mass of the carrier of that force: The larger the range, the smaller the mass of the carrier. Therefore, a force with an infinite range (the largest possible) will have a carrier with zero mass (the smallest possible). Indeed, the electromagnetic force has an infinite range, and the photon has zero mass. If the range of a force is known, there is a way to calculate or predict the mass of the carrier. A physicist, Hideki Yukawa, did this in the 1930s. From the range of the strong force, he calculated the mass of the carrier of the force to be 1/7 the mass of the proton and was also able to predict that it came in three charge varieties: positive, negative, and neutral. This particle was eventually named the **pion**. The charged pion was discovered in 1947 by Cecil Powell at Bristol University in England, and the first machine-produced pions were discovered in accelerator experiments in Berkeley in 1948 (charged ones) and in 1950 (neutral ones), with the masses that Yukawa had predicted. Now, as it turns out, pions are not the true carriers of the fundamental strong force between the parts of protons and neutrons (gluons are), but they do act in a sense as intermediary exchange particles between the nucleons. The important thing here is that Yukawa predicted the existence of a particle based on theory and that the particle was subsequently discovered in an experiment.

Pions are different from most of the other particles that we have encountered so far, in that they are unstable. An unstable particle lives for a short time and then spontaneously decays into other particles. The only other particle that we know that behaves this way is the neutron, when it undergoes radioactive beta decay into a proton, an electron, and a neutrino. The mean length of time that a particle exists before decaying is called the **lifetime** of the particle. The lifetime for the free neutron is about 15 minutes, which is extremely long compared to the lifetime of the pion. The pions with charge (+1 and -1) have a lifetime on the order of 10^{-8} seconds, and the neutral pion has a lifetime of about 10^{-16} seconds.

Let's summarize the pion particles.

- Pions come in three kinds. The most common decay mode is listed for each particle, but other decay modes are possible.

symbol	charge	mass	lifetime	decay mode
π^+	+1	1/7	10^{-8}	$\pi^+ \Rightarrow \mu^+ + \nu$
π^-	-1	1/7	10^{-8}	$\pi^- \Rightarrow \mu^- + \nu$
π^0	0	1/7	10^{-16}	$\pi^0 \Rightarrow \gamma + \gamma$

Table 3-1. Pion particle properties.

Muons

If you looked closely at the pion decay modes, you should have noticed some new symbols in the decay mode for the π^+ and the π^-. The symbols μ^+ and μ^- are for the particles called **muons**. In the search for pions, physicists discovered a particle with a mass of 1/9 the proton mass, not 1/7 as Yukawa had predicted. This was the muon. Now, it is known that pions can decay, as shown above, into muons, but when looking for the pions, ironically, they discovered the muons instead. Let's summarize the muon particles.

- Muons come in two kinds, and they are heavy (200 times as massive) "cousins" of the electron and positron. The most common decay mode is listed for each particle, but others are possible.

symbol	charge	mass	lifetime	decay mode
μ^+	+1	1/9	10^{-6}	$\mu^+ \Rightarrow e^+ + \nu + \nu$
μ^-	-1	1/9	10^{-6}	$\mu^- \Rightarrow e^- + \nu + \nu$

Table 3-2. Muon particle properties.

Three Kinds of Neutrinos and Antineutrinos

Actually, not all the decay modes in the last section are correct in the strictest sense. Usually a positive pion decays in the way mentioned above:

$$\pi^+ \Rightarrow \mu^+ + \nu$$

but sometimes it decays in the following fashion:

$$\pi^+ \Rightarrow e^+ + \nu$$

The problem here is that the neutrinos in the two reactions are different. The one produced in the decay to a muon is called a **muon neutrino**, and to show this difference we attach the subscript μ to the symbol. The one produced in the decay to an electron is called an **electron neutrino**, and we attach to it the subscript e. The correct expressions for the decays are:

$$\pi^+ \Rightarrow \mu^+ + \nu_\mu$$

$$\pi^+ \Rightarrow e^+ + \nu_e$$

If they both have no mass and no charge, how do we really know that they are different particles? If you could "follow" the two neutrinos produced in the reactions above, you would find something very interesting. If the muon neutrino hits a neutron, this will occur:

$$n + \nu_\mu \Rightarrow \mu^- + p$$

but this will not:

$$n + \nu_\mu \Rightarrow e^- + p$$

That is, you will never see a muon neutrino (the one produced in conjunction with the positive muon) interact with a neutron to produce a proton and an electron. It can only produce a proton and negative muon. As well, if the electron neutrino hits a neutron, this will occur:

$$n + \nu_e \Rightarrow e^- + p$$

but this will not:

$$n + \nu_e \Rightarrow \mu^- + p$$

That is, you will never see an electron neutrino (the one produced in conjunction with the positron) interact with a neutron to produce a proton and a muon. It can only produce a proton and an electron.

To complicate things even further, every particle must have an antiparticle, and neutrinos are no exception. Each of the neutrinos, therefore, has a corresponding antineutrino, symbolized with a bar on top.

$$\nu_e \qquad \overline{\nu}_e$$

$$\nu_\mu \qquad \overline{\nu}_\mu$$

Straying from our historical picture, so that we may complete the discussion of neutrinos, a heavier cousin of the electron and muon, the tau (τ), was discovered recently (1975). The tau can have positive or negative electric charge and has a mass of close to twice the proton. Of course, it has a neutrino and an antineutrino that belong with it. So, in all, there are three neutrinos and three antineutrinos.

Particle Classifications

The electron, muon, tau, their corresponding antiparticles, neutrinos, and antineutrinos are known collectively as **leptons**. Leptons are spin-1/2 particles that do not experience the strong force.

Particles that do experience the strong force are called **hadrons**. There are two kinds of hadrons: baryons and mesons. Hadrons with half-integer spin are known as **baryons** and those with integer spin are called **mesons**. So far, the only baryons we have seen are the proton and neutron, and the only meson we have encountered is the pion.

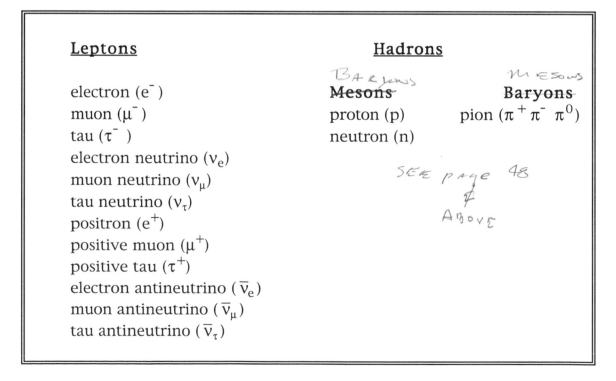

Leptons   Hadrons

BARYONS   *MESONS*

electron (e^-)   **Mesons**   **Baryons**
muon (μ^-)   proton (p)   pion ($\pi^+ \pi^- \pi^0$)
tau (τ^-)   neutron (n)
electron neutrino (ν_e)
muon neutrino (ν_μ)   *SEE page 48*
tau neutrino (ν_τ)
positron (e^+)   *↑*
positive muon (μ^+)   *ABOVE*
positive tau (τ^+)
electron antineutrino ($\overline{\nu}_e$)
muon antineutrino ($\overline{\nu}_\mu$)
tau antineutrino ($\overline{\nu}_\tau$)

Table 3-3. Particle classifications.

Summary

In this chapter:

We discussed some of the experiments and theories that led to discoveries of new particles.

- The antielectron or positron was predicted from theory and then subsequently discovered in an experiment. Antiparticles have opposite properties to their corresponding particles when an opposite exists. The antiparticles of the electron, proton, and neutron are:

    - the positron.
    - the antiproton.
    - the antineutron.

- Yukawa predicted a particle with 1/7 the mass of the proton to be the carrier of the strong force between nucleons. It was subsequently discovered as an unstable particle, the pion that comes in three charge varieties.

"The noble pions seem to be giving themselves up to a greater cause. Through the pions, protons and neutrons are getting together. Protons are laying down with protons, neutrons with neutrons together to form atoms! There there look for yourself. Atoms, I tell you, atoms! Somehow, through the pions jumping back and forth transmitting the strong force, the nuclei stay together. Millions of pions give themselves for us all everyday..." Greg Bruna

- The search for the pions and the study of their decays led to the discovery of more new particles:

    - muons.
    - two kinds of neutrinos (eventually expanded to three).

- We started to classify the particles according to common properties as:

    - leptons: particles that do not experience the strong force.
    - hadrons: particles that do experience the strong force, of which there are two kinds, mesons and baryons. Mesons have integer spins and baryons have half-integer spins.

Self-Test 3

For multiple choice, check all that apply.

1. The positron has the same spin and charge as

       a. an electron.
       b. a proton.
       c. a neutrino.

2. The antiproton has

       a. a positive charge.
       b. a negative charge.
       c. no charge.

3. The neutron and the antineutron have

       a. the same mass.
       b. little or no mass.
       c. the same spin.

4. The pion

       a. was predicted from theory before it was observed.
       b. is stable.
       c. can decay to a muon and a neutrino.
       d. is a lepton.

5. The muon

       a. was predicted from theory before it was observed.
       b. is stable.
       c. cannot decay to pions.
       d. is a lepton.

6. Using the rules that you have learned so far, indicate whether each of these particle processes can or cannot occur.

a) $\quad e^- + e^- \Rightarrow e^- + e^-$

b) $\quad e^- + e^+ \Rightarrow e^+ + e^-$

c) $\quad \mu^- + e^+ \Rightarrow \mu^- + e^+$

d) $\quad e^+ + e^- \Rightarrow$ energy

e) $\quad \mu^+ + e^- \Rightarrow$ energy

f) $\quad p + \bar{n} \Rightarrow$ energy

g) $\quad$ energy $\Rightarrow \bar{n} + n$

7. Fill in the particle review chart. Remember, all masses are approximate and should be quoted in terms of the proton mass.

particle	charge	mass	lifetime	feels strong force
electron		1/1800	stable	no
neutrino				
muon	+1/-1		10^{-6} sec	no
proton		1		
neutron			930 sec	
pion	+1/-1			yes
pion	0			yes

8. Fill in the review chart on antimatter

object	symbol	mass	charge	spin
antiproton				
antineutron				
antielectron				

9. Review the 12 leptons by filling in the chart.

object	symbol	mass	charge	matter or antimatter
electron		1/1800		matter
positron		1/1800		
muon		1/9	-1	
muon		1/9	+1	antimatter
tau		2	-1	
tau		2	+1	
	ν_e			
	$\bar{\nu}_e$			
	ν_μ			
	$\bar{\nu}_\mu$			
	ν_τ			
	$\bar{\nu}_\tau$			

Answers to Self-Test 3

1. The positron has the same spin and charge as

    a. an electron.
    b. a proton.   X
    c. a neutrino.

2. The antiproton has

    a. a positive charge.
    b. a negative charge. X
    c. no charge.

3. The neutron and the antineutron have

    a. the same mass.   X
    b. little or no mass.
    c. the same spin.   X

4. The pion

    a. was predicted from theory before it was observed.   X
    b. is stable.
    c. can decay to a muon and a neutrino.   X
    d. is a lepton.

5. The muon

    a. was predicted from theory before it was observed.
    b. is stable.
    c. cannot decay to pions.   X
    d. is a lepton.   X

6. Processes that can and cannot occur:

a) $e^- + e^- \Rightarrow e^- + e^-$ (yes)

b) $e^- + e^+ \Rightarrow e^+ + e^-$ (yes)

c) $\mu^- + e^+ \Rightarrow \mu^- + e^+$ (yes)

d) $e^+ + e^- \Rightarrow$ energy   (yes)

e) $\mu^+ + e^- \Rightarrow$ energy   (no)

f) $p + \bar{n} \Rightarrow$ energy   (no)

g)   energy $\Rightarrow \bar{n} + n$ (yes)

If you do not understand why e and f cannot occur, refer to Pages 30-31.

7. Particle review chart

particle	charge	mass	lifetime	feels strong force
electron	-1	1/1800	stable	no
neutrino	0	0	stable	no
muon	+1/-1	1/9	10^{-6} sec	no
proton	+1	1	stable	yes
neutron	0	1	930 sec	yes
pion	+1/-1	1/7	10^{-8} sec	yes
pion	0	1/7	10^{-16} sec	yes

8. Answers to the review chart on antimatter

object	symbol	mass	charge	spin
antiproton	$\bar{p}$	1	-1	1/2
antineutron	$\bar{n}$	1	0	1/2
antielectron	e^{+}	1/1800	+1	1/2

9. Answers to lepton chart

object	symbol	mass	charge	matter/antimatter
electron	e^-	1/1800	-1	matter
positron	e^+	1/1800	+1	antimatter
muon	μ^-	1/9	-1	matter
muon	μ^+	1/9	+1	antimatter
tau	τ^-	2	-1	matter
tau	τ^+	2	+1	antimatter
electron neutrino	ν_e	0	0	matter
e antineutrino	$\overline{\nu}_e$	0	0	antimatter
muon neutrino	ν_μ	0	0	matter
muon antineutrino	$\overline{\nu}_\mu$	0	0	antimatter
tau neutrino	ν_τ	0	0	matter
tau antineutrino	$\overline{\nu}_\tau$	0	0	antimatter

Chapter 4 MORE PARTICLES AND CONSERVATION RULES

Most of the particle discoveries we have examined so far were done in the laboratory without the use of particle accelerators. As particle accelerators came into greater use in the 1950s, all sorts of new particles were discovered and studied. Understanding how these particles were produced and how they interacted with other particles was of crucial importance during that decade. In trying to devise theories that could explain why certain reactions occurred and others did not, physicists discovered new particle properties and defined conservation rules for these properties. At first it will seem as if these properties are merely invented mathematical concepts, but as we proceed we will discover that they have as much physical significance as charge, mass, and spin. In this chapter we will learn:

- About strange particles.

- More rules that govern the ways that particles interact.

- About additional particle properties:

  - lepton number.
  - baryon number.
  - strangeness.

- More about neutrinos.

Strange Particles

By the 1950s, particle accelerators were being used for more experiments and all sorts of new particles were discovered. These particles were produced from reactions between protons, neutrons, and pions; that is they were created during collisions between these particles. These particles were collectively called **strange particles** because there were some peculiarities in their production and decay:

- They were always produced in pairs.

- Though they were all produced from reactions containing protons and neutrons, some of them would decay and not have a proton or neutron as an end product.

- All of the strange particles are unstable.

- All of the strange particles are created by strong force interactions and often decay via the weak force.

Below is a listing of some strange particles. Where a decay mode is listed, it is the most common one. The lifetimes given are approximate.

family name = KAONS

mass = 1/2 proton mass

lifetime = 10^{-8} or 10^{-10} seconds

spin = 0

two particles K^+, K^0

two antiparticles K^-, $\overline{K}^0$

decay example

$$K^+ \Rightarrow \mu^+ + \nu_\mu$$

family name = LAMBDA

mass = 1.1 times the proton mass

lifetime = 10^{-10} seconds

spin = 1/2

one particle Λ

one antiparticle $\overline{\Lambda}$

decay example:

$$\Lambda \Rightarrow \pi^- + p$$

family name = SIGMAS

mass = 1.2 times the proton mass

lifetime = 10^{-10} or 10^{-20} seconds

spin = 1/2

three particles Σ^-, Σ^+, Σ^0

three antiparticles $\overline{\Sigma}^-$, $\overline{\Sigma}^+$, $\overline{\Sigma}^0$

decay examples:

$$\Sigma^+ \Rightarrow p + \pi^0$$
$$\Sigma^0 \Rightarrow \Lambda + \gamma$$
$$\Sigma^- \Rightarrow n + \pi^-$$

"Hi! I'm a negatively charged kaon, one of the most incredible subatomic particles in the universe, if I do say so myself. I'm making this tape at Planck's Particle Dating Service because I would like to meet another nice subatomic particle as soon as possible. I know I sound like I'm rushing, but when you only live for 10^{-8} seconds, you can't afford to waste time. Let me tell you about myself. I have half the mass of a proton. My mother was a neutron and my father was a pion. They collided and reacted strongly in a bubble chamber very recently -- they work fast..." John P. McCormick

family name = CASCADES

mass = 1.3 times the proton mass

lifetime = 10^{-10} seconds

spin = 1/2

two particles $\quad \Xi^0$, Ξ^-

two antiparticles $\overline{\Xi}^0$, $\overline{\Xi}^-$

decay examples:

$$\Xi^0 \Rightarrow \Lambda + \pi^0$$

$$\Xi^- \Rightarrow \Lambda + \pi^-$$

Reaction Rules

Review of the Conservation Rules

We have already seen some of the rules we can use to determine whether a reaction can take place. Let's review them before we go on to learn more rules. The pages where the rules are discussed in more detail are indicated.

- Charge Conservation: The total charge of the particles must be the same before and after the reaction. (Page 8)

- Energy Conservation for decays: The mass of the decaying particle must be greater than the sum of the masses of the end products. Remember, energy is conserved or balanced because the particles also have kinetic energy. (Page 9)

- Momentum Conservation: This is one rule that we will not be able to test ourselves, so we will assume that it is always obeyed. (Page 9)

- The neutrino type must match the lepton type. (Pages 33-34)

Neutrino/Antineutrino Rule

The first new rule that we will learn is related to neutrinos and antineutrinos. To illustrate this rule, let's go back to the reaction that gave us the concept of neutrinos in the first place. Remember neutron beta decay:

$$n \quad \Rightarrow \quad p + e^- + \nu$$

Well, this is not exactly correct. First of all, the ν must be a ν_e (neutrino type must match the lepton type), and it is actually an antineutrino. It may be difficult to understand why we know that it is an antineutrino and not a neutrino, when there is no apparent way to tell them apart. The reasons for this are complex and beyond the scope of this book, so I will just give you the rule that allows you to figure out which one it will be, neutrino or antineutrino, in a given reaction. When they are on the same side of a particle reaction or decay:

- Electrons must be accompanied by antineutrinos.
- Positrons must be accompanied by neutrinos.
- Negative muons must be accompanied by antineutrinos.
- Positive muons must be accompanied by neutrinos.

When they are on opposite sides of a particle reaction or decay:

- Electrons must be accompanied by neutrinos.
- Positrons must be accompanied by antineutrinos.
- Negative muons must be accompanied by neutrinos.
- Positive muons must be accompanied by antineutrinos.

The way physicists usually understand the rules above is in terms of yet another quantity that must be conserved. Originally termed **lepton number** and later expanded to include all leptons, the quantity is actually three quantities that must be separately conserved. These three quantities are called **electron lepton number**, **muon lepton number**, and **tau lepton number**, and the values for the leptons are shown in the table below. All other particles have lepton numbers of zero.

particle	electron lepton number	muon lepton number	tau lepton number
electron	+1	0	0
positron	-1	0	0
electron neutrino	+1	0	0
electron antineutrino	-1	0	0
negative muon	0	+1	0
positive muon	0	-1	0
muon neutrino	0	+1	0
muon antineutrino	0	-1	0
negative tau	0	0	+1
positive tau	0	0	-1
tau neutrino	0	0	+1
tau antineutrino	0	0	-1

Table 4-1. Lepton number values.

Some examples of proper reactions are:

$$n \Rightarrow p + e^- + \overline{v}_e$$

electron lepton number   0 = 0   +1   -1   conserved

$$n + \overline{v}_e \Rightarrow p + e^-$$

electron lepton number   0   +1 = 0   +1   conserved

Some reactions that violate lepton number conservation and therefore cannot take place are:

$$\mu^- \Rightarrow e^- + \gamma$$

electron lepton number	0	+1	0	not conserved
muon lepton number	+1	0	0	not conserved

$$n + v_e \Rightarrow p + \mu^-$$

electron lepton number	0	+1	0	0	not conserved
muon lepton number	0	0	0	+1	not conserved

Baryon Number

The second new rule has to do with the number of protons, neutrons, and other particles that belong to the baryon classification. The total number of baryons must remain constant. To simplify this, all baryons are assigned a baryon number of 1 and all nonbaryons (leptons and mesons) are assigned a baryon number of 0. An antiparticle has an opposite baryon number from its particle. To determine if a reaction can or cannot occur, compare the total baryon number of the incoming particles to the total baryon number of the outgoing particles.

p, n, Λ, Σ, Ξ have B = 1

π, K, e, μ, v, τ have B = 0

Let's apply our rules to see if certain processes can or cannot occur.

$$n \Rightarrow \pi^+ + \pi^-$$

charge	0	+1	-1	conserved
baryon	1	0	0	not conserved
mass	1	1/7	1/7	conserved

The decay above will not occur since baryon number is not conserved.

$$\pi^- + p \Rightarrow K^+ + \Sigma^-$$

charge	-1	+1	+1	-1	conserved
baryon	0	1	0	1	conserved

The reaction above should occur and indeed is observed.

Strangeness

Next consider a reaction that differs only slightly from the previous example.

$$\pi^- + p \Rightarrow \pi^- + \Sigma^+$$

charge	-1	+1	-1	+1	conserved
baryon	0	1	0	1	conserved

The reaction above should occur since it passes all our known rules. However, it has never been observed in an experiment. Comparing this reaction to the previous one, we can see that, if anything, the second reaction should occur more frequently than the first. The reason for this is that it is easier to create a pion and a sigma than it is to create a kaon and a sigma because of the mass difference between the pion and the kaon (the kaon is three times as massive as the pion). So how do we explain why the second reaction does not occur? The answer is simple. It does not occur because it is in violation of some rule that we do not know yet. This is essentially the same reasoning the physicists had in the 1950s, when they were studying these reactions. The rule that they proposed involved a property they called **strangeness**. It is another property that a particle may or may not have, just like charge, mass, spin, and baryon number. It must be conserved in all strong and electromagnetic interactions. If strangeness is not conserved, the reaction may still occur under the influence of the weak force. Since neutrinos are not influenced by the strong or electromagnetic force, you do not have to worry about strangeness conservation for any reaction with neutrinos. The values of strangeness that a particle can have are +1, 0, -1, -2 or -3. (We will understand more deeply the reasons for these particular values in Chapter 5.)

The assignment of the strangeness values begins with pions, protons, and neutrons, all of which have strangeness (S) equal to zero. If they are the only objects in the initial state, they may produce only particles whose total strangeness is zero. For example:

$$\pi^- + p \Rightarrow n + \pi^0$$

But all the other hadrons have non-zero strangeness. We can determine what the values are by arbitrarily (just as the physicists did) assigning a strangeness of +1 to the K$^+$ and looking at some reactions that occur. For example, the following reaction does occur, and it is a strong interaction:

$$n + \pi^0 \Rightarrow \Sigma^- + K^+$$

	n	π^0	Σ^-	K^+
charge	0	0	-1	+1
baryon	1	0	1	0
strangeness	0	0	?	+1

It tells us that the Σ^- has a strangeness of -1. We can proceed in this manner to assign a value of strangeness to all the particles we know of so far. The results are shown in the table below.

name	mass	strangeness
p	1	0
n	1	0
Λ	1.1	-1
Σ	1.2	-1
Ξ	1.3	-2
π	1/7	0
K^+, K^0	1/2	+1

Table 4-2. Strangeness values.

All the antiparticles have the opposite strangeness of the particles listed above.

50

Summary

In this chapter:

- We learned about several new types of particles.

    - kaons, mesons with a mass of 0.5
    - lambda, baryon with a mass of 1.1
    - sigmas, baryons with a mass of 1.2
    - cascades, baryons with a mass of 1.3

- We reviewed some old rules and learned some new ones for determining whether or not reactions take place. They are:

    - Charge conservation.

    - Baryon number conservation.

    - Strangeness conservation.

    - Lepton and neutrino (antineutrino) matching.

    - The mass of the decaying particle must be greater than the mass of the end products.

- Good things to remember:
    - Any interaction with a neutrino in it must be weak.

    - Any interaction with a photon (γ) in it must be electromagnetic.

    - If strangeness is conserved, the reaction *will* occur via the strong or electromagnetic force.

    - If strangeness is not conserved, the reaction still *may* occur through the weak force.

Self-Test 4

For multiple choice, check all that apply.

1. If the baryon number of a proton is 1, the baryon number of the antiproton is

     a.   1
     b.   0
     c. -1

2. The baryon number of an electron is

     a.   1
     b.   0
     c. -1

3. Strangeness must be conserved in

     a. weak interactions.
     b. strong interactions.
     c. electromagnetic interactions.

4. Strange particles

     a. do not decay.
     b. decay very quickly.
     c. are leptons.

5. Indicate whether or not the following reactions can take place. If they can occur, then indicate the force responsible. You may need to look back to the chapter summary or even previous chapters for some of the property values. Remember, just as antiparticles have electric charge equal but opposite to their corresponding particles, they have equal but opposite strangeness and baryon number.

$$p \, \bar{p} \Rightarrow \pi^+ \, \pi^- \, \pi^- \, \pi^+ \, \pi^0$$

$$p \, K^- \Rightarrow \Sigma^+ \, \pi^- \, \pi^- \, \pi^+ \, \pi^0$$

$$p \, \pi^- \Rightarrow \Lambda \, \bar{\Sigma}^0$$

$$p \, \bar{\nu}_\mu \Rightarrow \mu^+ \, n$$

$$p \, \bar{\nu}_\mu \Rightarrow e^+ \, n$$

$$p \, \nu_e \Rightarrow e^+ \, \Lambda \, K^0$$

$$p \, \nu_e \Rightarrow e^- \, \Sigma^+ \, K^+$$

$$e^+ \Rightarrow \mu^+ \, \bar{\nu}_e \, \nu_\mu$$

$$K^+ \Rightarrow \pi^+ \, \pi^0$$

$$\Lambda \Rightarrow p \, K^-$$

1. If the baryon number of a proton is 1, the baryon number of the antiproton is

     a.   1
     b.   0
     c. -1   X

2. The baryon number of an electron is

     a.   1
     b.   0   X
     c. -1

3. Strangeness must be conserved in

     a. weak interactions.
     b. strong interactions.   X
     c. electromagnetic interactions.   X

4. Strange particles

     a. do not decay.
     b. decay very quickly.   X
     c. are leptons.

5. In the following, C = charge conservation, B = baryon number conservation, S = conservation of strangeness, L = lepton neutrino/antineutrino match, E = energy conservation for decays. A check mark indicates that the process upholds the conservation rule and an X indicates that the process violates the rule. NA indicates that that particular rule is not applicable.

	C	B	S	L	E	
$p\,\bar{p} \Rightarrow \pi^+\,\pi^-\,\pi^-\,\pi^+\,\pi^0$	√	√	√	NA	NA	yes strong

Everything that needs to be conserved is, and therefore this reaction can take place via the strong force.

	C	B	S	L	E	
$p\,K^- \Rightarrow \Sigma^+\,\pi^-\,\pi^-\,\pi^+\,\pi^0$	√	√	√	NA	NA	yes strong

Everything is conserved in this case, too. If you were concerned that the masses of the final particles might be more than the masses of the initial particles, don't be. In this reaction, the initial particles can bring kinetic energy into the process to make up any additional mass.

	C	B	S	L	E	
$p\,\pi^- \Rightarrow \Lambda\,\bar{\Sigma}^0$	√	X				no

Baryon number is not conserved here.

	C	B	S	L	E	
$p\,\bar{\nu}_\mu \Rightarrow \mu^+\,n$	√	√	√	√	NA	yes weak

We know that this must be the weak force because of the presence of the neutrino. Remember, neutrinos are only affected by the weak force.

	C	B	S	L	E	
$p\,\bar{\nu}_\mu \Rightarrow e^+\,n$	√	√	√	X	NA	no

This cannot occur because the neutrino type doesn't match the lepton type.

	C	B	S	L	E	
$p\,\nu_e \Rightarrow e^+\,\Lambda\,K^0$	√	√	NA	X	NA	no

This cannot occur because electron lepton number is not conserved.

	C	B	S	L	E	

$p\,\nu_e \Rightarrow e^- \ \Sigma^+ \ K^+$ √   √   √   √   NA   yes weak

This must be the weak force because of the presence of the neutrino.

$e^+ \Rightarrow \mu^+ \ \overline{\nu}_e \ \nu_\mu$ √   NA   NA   √   X   no

Since this reaction is a decay, the mass of the initial particle must be greater than the sum of the masses of the final particles. The mass of the electron is much less than the mass of the muon, and therefore the reaction cannot take place.

$K^+ \Rightarrow \pi^+ \ \pi^0$ √   NA   X   NA   √   yes weak

Even though strangeness is not conserved, this reaction can take place via the weak force because the weak force does not have to conserve strangeness.

$\Lambda \Rightarrow p \ K^-$ √   √   √   NA   X   no

Since this reaction is a decay, the mass of the initial particle must be greater than the sum of the masses of the final particles for it to occur. The mass of the lambda is less than the mass of the proton and kaon combined.

Chapter 5 SIMPLIFICATION OF THE ZOO: QUARKS

The view of matter that had evolved by the late 1950s was not a simple one. Physicists had taken the three-particle model of protons, neutrons, and electrons and expanded it to include a very large number of fundamental entities. They began to ask tougher questions:

- Are all these particles really fundamental, or are they, too, composed of parts?
- If these parts exist, what are they like and how many are there?
- Where is the underlying simplicity expected from nature?

Physicists devised theories to understand the complex world of subatomic particles. One theory proposed that the proton and the neutron, which make up the nucleus, were made of a more basic particle. The new particle was named the **quark**. The quark theory could explain all the hadrons in Table 5.4 (page 61), but experimental evidence was needed to support the theory. On completion of this chapter you will be able to:

- List six quarks.
- Combine quarks and antiquarks in pairs to form mesons.
- Combine three quarks to form baryons.
- Explain the early experiments supporting the theory of quarks.

The Quark Model

As early as 1964, Murray Gell-Mann and George Zweig independently came up with a theory that would explain all the hadrons that we have discussed so far. Theirs was a model based on three constituents, all with spin of 1/2 and baryon number of 1/3. Gell-Mann named these new particles quarks. Quarks come in three types (called **flavors** by physicists): up, down, and strange. By convention, each strange quark contributes a strangeness of -1, and each antistrange quark contributes a strangeness of +1.

name	symbol	charge	strangeness
up	u	+2/3	0
down	d	-1/3	0
strange	s	-1/3	-1
antiup	$\bar{u}$	-2/3	0
antidown	$\bar{d}$	+1/3	0
antistrange	$\bar{s}$	+1/3	+1

Table 5-1. Quark properties.

Quark/Antiquark Combinations (Mesons)

We can make mesons out of quark/antiquark combinations. That is, if we take all the possible combinations of a quark and an antiquark and figure out the properties of each combination, we will have all the mesons we have discussed and more. The following table shows all these combinations. You should refer to the review table on hadrons (Table 5-4) as you go through these. Since each meson can contain either one strange quark, one antistrange quark, or one of each, the only possible values for the strangeness of mesons are -1, +1, and 0. Notice too that there are three particles listed as being composed of an up and an antiup quark. These are indeed different particles, as the quarks are oriented differently in each case. We will see this in more detail later in this chapter. Since the 1960s when the quark theory was first proposed, many, many mesons have been discovered, and all of them are composed of a quark and an antiquark. No meson has been found that does not fit this model.

combination	charge	strangeness	particle
$u\bar{u}$	0	0	$\pi^0, \eta^0, \eta^{0\prime}$
$\bar{d}d$	0	0	$\pi^0, \eta^0, \eta^{0\prime}$
$\bar{u}d$	-1	0	π^-
$u\bar{d}$	+1	0	π^+
$u\bar{s}$	+1	+1	K^+
$\bar{u}s$	-1	-1	K^-
$d\bar{s}$	0	+1	K^0
$\bar{d}s$	0	-1	$\bar{K}^0$
$s\bar{s}$	0	0	$\eta^{0\prime}$

Table 5-2. Quark/antiquark combinations.

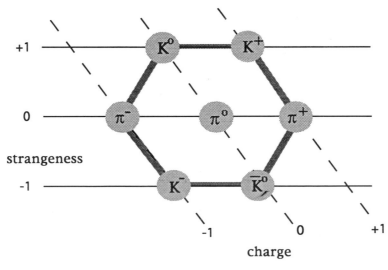

Figure 5-1. Mesons.

Three-Quark Combinations (Baryons)

We can make all the baryons out of three-quark combinations. That is, if we take all the possible combinations of three quarks and figure out the properties of each combination, we will have all of our baryons (and more). The following table shows all these combinations. You should refer again to the review table on hadrons (Table 5-4) as you go through these. Note that there are some new particles on the list. They are distinguishable by their masses and spin values. The more-familiar particles in column one all have spin of 1/2, and the second column of particles all have spin of 3/2. As you go down either column, the particles get more and more massive. Additional baryons have been discovered since the quark model was proposed, and all can be considered to be composed of three quarks. Antibaryons, like the antiproton, are all composed of three antiquarks. Now we can see that since baryons can contain 1, 2, or 3 strange quarks, the possible strangeness values for baryons are -1, -2, and -3. For the antibaryons the possible values of strangeness are 1, 2, and 3.

combination	charge	strangeness	particle	
uud	1	0	p	Δ^+
udd	0	0	n	Δ^0
uuu	2	0		Δ^{++}
ddd	-1	0		Δ^-
uus	1	-1	Σ^+	Σ^{+*}
uds	0	-1	Λ, Σ^0	Σ^{0*}
dds	-1	-1	Σ^-	Σ^{-*}
uss	0	-2	Ξ^0	Ξ^{0*}
dss	-1	-2	Ξ^-	Ξ^{-*}
sss	-1	-3		Ω^-

Table 5-3. Three quark combinations.

On Page 58, the arrangement of mesons formed a pattern. When we arrange the spin 1/2 baryons by charge and strangeness, curiously we get the same pattern. These patterns and the one for the spin 3/2 baryons were part of a theory called the Eightfold Way put forth in 1961. Arranging things this way led to further understanding of the quark model and the prediction of a new particle, the Ω, which was subsequently discovered at Brookhaven National Laboratory in 1963.

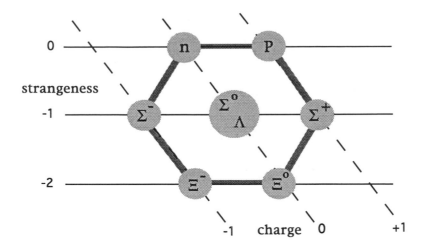

Figure 5-2. Baryons (spin 1/2).

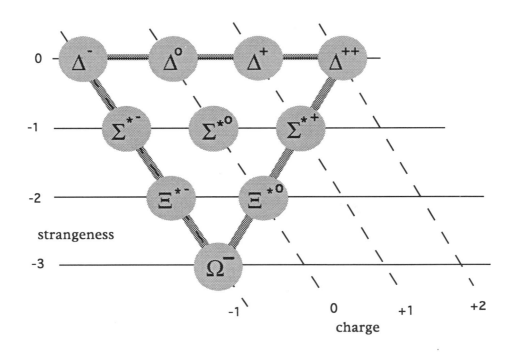

Figure 5-3. Baryons (spin 3/2).

name	charge	mass	strangeness	lifetime
proton (p)	+1	1	0	$> 10^{30}$ yrs
neutron (n)	0	1	0	930 s
lambda (Λ)	0	1.1	-1	10^{-10} s
sigma (Σ)	+1, -1	1.2	-1	10^{-10} s
sigma (Σ)	0	1.2	-1	10^{-20} s
cascade (Ξ)	0, -1	1.3	-2	10^{-10} s
pion (π)	+1, -1	1/7	0	10^{-8} s
pion (π)	0	1/7	0	10^{-16} s
kaon (K)	+1, -1	1/2	+1, -1	10^{-10} s
kaon (K)	0	1/2	+1	10^{-10} s
kaon ($\overline{K}$)	0	1/2	-1	10^{-10} s

Table 5-4. Hadron properties.

Antiparticles and Spin Considerations

An antiparticle contains the antiquarks that correspond to the quarks in the particle. Some examples of particle and antiparticle pairs are:

---

the proton and antiproton

$$p = uud \qquad\qquad \bar{p} = \bar{u}\,\bar{u}\,\bar{d}$$

---

the neutron and the antineutron

$$n = udd \qquad\qquad \bar{n} = \bar{u}\,\bar{d}\,\bar{d}$$

As promised back in Chapter 3, we can now see that, although the neutron and antineutron appear the same from the outside, with the same mass, charge, and spin, they are quite different on the inside. The neutron contains three objects, each with the same mass, but with charges +2/3, –1/3, and –1/3. The antineutron also contains three objects with the same mass; however, the charges are –2/3, +1/3, and +1/3. The neutron and the antineutron are indeed *different* objects.

---

the positive pion and the negative pion

$$\pi^{+} = u\bar{d} \qquad\qquad \pi^{-} = \bar{u}\,d$$

The positive pion is the antiparticle of the negative pion, and the negative pion is the antiparticle of the positive pion.

---

the neutral pion is its own antiparticle

$$\pi^{0} = u\bar{u} \text{ or } d\bar{d}$$
$$\text{anti}\,(\pi^{0}) = \bar{u}u \text{ or } \bar{d}d$$

Since the order does not matter, the neutral pion and the anti–neutral pion are the same particle.

But, how then, do we tell the difference between the Λ and the neutral Σ, when they are both made of the same three quarks? The answer lies in another property that we have already discussed: spin. Figure 5-4 shows what the internal spin states of the quarks are for the Λ and the neutral Σ. The Λ has the up quark and the down quark spinning opposite, with the strange quark spinning either the same as the up or the same as the down. The total spin is still 1/2. The Σ has the up and down quarks spinning the same way and the strange quark spinning opposite. This configuration has a higher energy than the Λ, hence the mass of the Σ is higher than the Λ. We can also understand the mass difference between the proton (udu) and the lambda (uds), since the strange quark is more massive than the up and down quarks.

mass of up quark   1/3 proton mass
mass of down quark   1/3 proton mass
mass of strange quark   1/2 proton mass

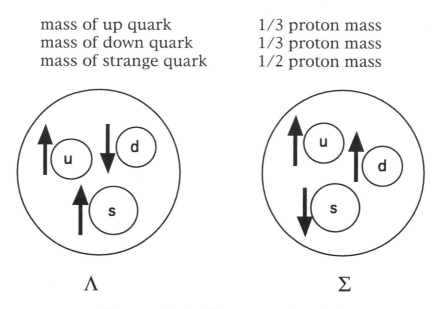

Λ   Σ

Figure 5-4. Sigma vs. lambda.

There is one more property of quarks that you should know about. The different flavors (u, d, s) come in colors (red, green, blue). These are not colors as we know color though, it is just a property that quarks possess. The reason being the original proposal for a color theory came about because of the existence of the Ω particle. You see, quarks have spin of 1/2 unit either up or down, i.e. there are two possible spin conditions for a given quark. Particles with 1/2 unit of spin that are identical (in every way) cannot exist in the same entity (i.e. two identical electrons cannot exist in one atom and two identical quarks cannot exist in the same baryon). The Ω has three strange quarks, identical in terms of mass, spin orientation (either all three are up or all three are down), flavor, and charge. This is forbidden by an important physics principle called the Pauli Exclusion Principle. The only way for the Ω to exist is if there is some other property that distinguishes the three strange quarks from each other. This is the property we call color. The very successful theory of color is called quantum chromodynamics.

Experimental Evidence for Quarks

The model works, but are quarks real? Has a quark ever been seen? The answer is yes and no. No experiment has ever detected a *free* quark. But there is compelling evidence that quarks exist *inside* protons, neutrons, and other particles. In the 1970s, electrons and neutrinos were used to probe protons just as the alpha particles were used to probe the atom and the nucleus in the early part of the century. Figure 5-5 shows the electrons scattering off a proton made up of quarks. The results here were analogous to Rutherford's experiments. The electrons emerged from the reaction at angles consistent with the quark model of the proton. These experiments were able to show that the proton was composed of three objects, each with spin of 1/2 and two with charges of + 2/3 and one with a charge of -1/3. Later versions of this same type of experiment also gave evidence for the existence of gluons inside the proton, supporting the theory that they are the carriers of the strong force. There have been searches made for free quarks, which should be fairly easy to detect due to their unusual fractional electric charges, but with no success to date. The present theory supports the existence of quarks, but as objects permanently confined inside hadrons. That is, we may never see a free quark. What about color? Can we ever see the color of a quark? The answer is not directly. However, experiments were conducted in the early 1970s in Italy and the United States which confirmed that quarks have color.

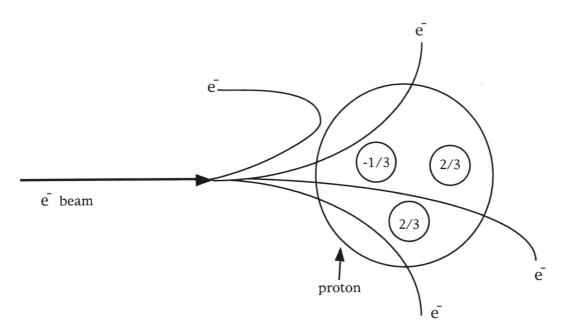

Figure 5-5. Electron-proton scattering.

Summary

In this chapter:

- We learned about three quarks.

    - up: with electric charge +2/3
    - down: with electric charge -1/3
    - strange: with electric charge -1/3

- We saw how all the known mesons can be made from the combination of a quark and an antiquark.

- We saw how all known baryons can be made from combinations of three quarks.

- We discussed the evidence that supports the existence of quarks.

"A typical day in the life of the Atom's family starts off with mom-protons and dad-neutrons waking up. Both mom and dad care for the kid-quarks. Basically caring for the kid quarks means making sure they are always changing their quark colors and making sure they are having fun with the many games that they play, like 'spin' (which is a game they always play), where all they do is constantly spin around! Everybody in the family must obey simple house rules like the Uncle Pauli exclusion principle (which says that there can't be more than one person with the same color, spin and state in the same room). Mom-proton and dad-neutron are very protective of the kid-quarks! They never let the kid-quarks wander around alone - if this should one day happen, mom-proton and dad-neutron would literally 'fall apart'." Shireen Kaufman

Self-Test 5

For multiple choice, check all that apply.

1. The quark theory was proposed to

        a. simplify our view of matter.
        b. explain all the new particles that were discovered in the 70s.
        c. explain spin.

2. Quarks and antiquarks have

        a. no spin
        b. integer spins
        c. spin of 1/2

3. The following quarks have non-zero strangeness

        a. up
        b. down
        c. strange
        d. antistrange

4. Which particle has the same exact quarks as a neutron?

        a. the proton
        b. the antineutron
        c. the neutral pion
        d. none of the above

5. There is

        a. no experimental evidence that quarks exist.
        b. experimental evidence for free quarks.
        c. experimental evidence that quarks exist inside particles.

6. Give the quark content of the following. Are any of them particle/antiparticle pairs?

π^+ $\qquad\qquad$ π^- $\qquad\qquad$ π^0 $\qquad\qquad$ K^+

K^- $\qquad\qquad$ Λ $\qquad\qquad$ Σ^0

7. Write down the equation for neutron beta decay.

Explain this decay in terms of quarks and draw the Feynman diagram.

8. Draw a picture of the helium atom as seen by a physicist in 1970.

Answers to Self-Test 5

1. The quark theory was proposed to

    a. simplify our view of matter.   X
    b. explain all the new particles that were discovered in the 70s.
    c. explain spin.

2. Quarks and antiquarks have

    a. no spin
    b. integer spins
    c. spin of 1/2   X

3. The following quarks have non-zero strangeness

    a. up
    b. down
    c. strange   X
    d. antistrange   X

4. Which particle has the same exact quarks as a neutron?

    a. the proton
    b. the antineutron
    c. the neutral pion
    d. none of the above   X

5. There is

    a. no experimental evidence that quarks exist.
    b. experimental evidence for free quarks.
    c. experimental evidence that quarks exist inside particles.   X

6. Quark content: $(\pi^+ \pi^-)$ and $(K^+ K^-)$ are particle and antiparticle, and the π^0 is its own antiparticle.

π^+	π^-	π^0	K^+
$u\bar{d}$	$\bar{u}d$	$u\bar{u} + \bar{d}d$	$u\bar{s}$

K^-	Λ	Σ^0
$\bar{u}s$	uds	uds

7. Neutron beta decay equation

$$n \Rightarrow p + e^- + \bar{v}_e$$

Quark model:

$$\begin{matrix} u & & u \\ d & & d \end{matrix}$$

$$d \Rightarrow u + e^- + \bar{v}_e$$

Feynman diagram:

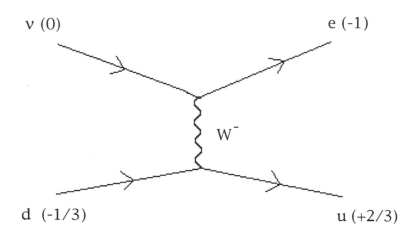

$v\,(0)$ $e\,(-1)$

W^-

$d\ (-1/3)$ $u\ (+2/3)$

69

8. A picture of the helium nucleus as seen by a physicist in 1970 (not to scale).

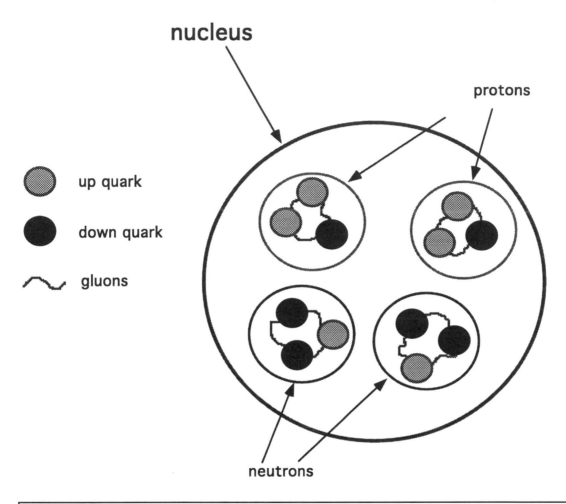

nucleus

protons

up quark

down quark

gluons

neutrons

"Once upon a time, in the land of Subatomia, there was a tiny city called Nucleus. In this city lived the ancient and mysterious Quarks. What are Quarks? Why, I'll tell you! These creatures were said to have their ups and downs. You see, no one had ever actually seen the wondrous Quarks. They were protected from the eyes of the world by the Gluons, a strong force of warriors who held all of the Quarks together. Yes, the Gluons were very important people in the Nucleus, and although the Neutrons had little to say in the matter, the Protons frequently proclaimed that the hard work of the Gluons kept the entire city together! All in all the Nucleus was a very positive city. Now outside the wall of Nucleus lived a diligent but lonely little electron named Leo. One day he was experiencing an existential dilemma. " What am I doing here? I am useless! The Gluons do all the work and I just go around and around and around…" Erica Bailey Carpenter

Chapter 6 THE STANDARD MODEL

The quark theory of the early 1960s is not the model that particle physicists accept. That model was only a beginning and we will see in this chapter how the experiments of the 1970s and '80s led to the current standard model of quarks and leptons. The quark theory also allows us to go back and get a deeper understanding of some concepts that we have learned. In this chapter we will:

- See the fundamental difference between the weak and strong forces.
- Complete the quark model, which today consists of six quarks to go with the six leptons.

The Strong and Weak Forces Revisited

The strong force simply rearranges quarks or creates quark/antiquark pairs from other quark/antiquark pairs. It *cannot* change one type of quark into another type of quark. Figure 6-1 is an example of a process that is governed by the strong force and a diagram showing the regrouping of quarks in that process.

$$\pi^- \; + \; p \; \Rightarrow \; n \; + \; \pi^0$$

$$\bar{u}d \qquad uud \qquad udd \qquad \bar{u}u$$

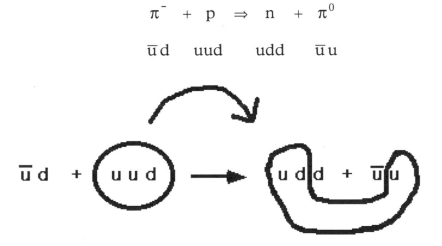

Figure 6-1. Quark regrouping.

"We met in the Cafe de Carbon. I was but a proton then working extra hours to make a living, and she, she was a negative pion, the most enchanting particle you have ever seen. Her charge electrified me, and we were drawn together with irresistible force. We danced the moment away, and the reaction was inevitable: we were married. It was unbelievable at first, the chemistry was all there. A rho particle made a bid for her, but my pull was too strong: we became one. As the instant rolled by, though, something happened. She had lost something, and the electricity, the vital spark of life was all gone. I had lost mine too, I guess, and I had put on a little weight. " You've become so... so... neutral," she sobbed, "you're just a big fat neutron! I remember the sting of the words..." James Chapin

Figure 6-2 is another example of a strong force at work. In this case, we see regrouping, but we also see that an up and antiup annihilated into energy, which then produced a strange/antistrange quark pair.

$$\pi^- \; + \; p \; \Rightarrow \; \Lambda \; + \; K^0$$

$$\bar{u}d \qquad uud \qquad uds \qquad \bar{s}d$$

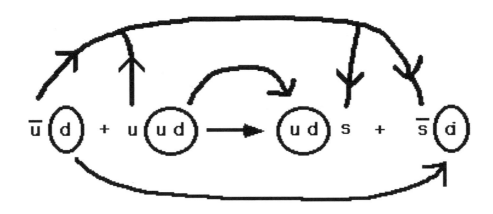

Figure 6-2. Quark regrouping and pair-production.

The weak force on the other hand, *can* change one type of quark into another type of quark, as shown in the example below. The strange quark in the Λ changes into an up quark and in so doing emits a W⁻ (a carrier of the weak force), which subsequently becomes a down quark and an antiup quark.

$$\Lambda \quad \Rightarrow \quad p \ + \ \pi^-$$

$$uds \qquad uud \quad \bar{u}d$$

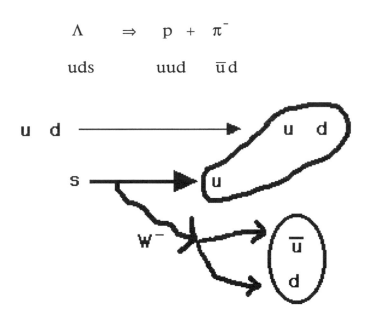

Figure 6-3. Quark transformation.

"I am writing this because I feel like something might be happening in me that I can't explain to anyone else. I've heard of this sort of thing occurring, but nobody likes to talk about it. I'm a neutron living in Carbon Atom 0-9-0-2000. We're a relatively peaceful, stable community of particles. I grew up with seven other neutrons and six protons. We were preached to about how the Strong Force governs us, affects us and controls us -- that we should be thankful for it... But the thing is that lately I've been having these other feelings. I've heard of the Weak Force in certain circles, in hushed voices... I started tracking down rare sources of information on the Weak Force. The ancient scrolls revealed that the Weak Force is also quite real and present in our world. Not only does it continuously change the flavor of the quarks within us, changing and affecting our moods, and interactions with others, but the Weak Force has the power to transform... The Weak Force isn't bad. I can feel that. The mythical W's and Z's of frightening fairy tales, presumed to be more massive than imaginable, who carry the Weak Force with them, have become a reality on my mind. I am in fact looking for them, awaiting them, awaiting a change..." Lori Lamont

For more practice with quarks, go back to Page 53 and check those reactions in terms of quarks.

The Standard Model

Up, Down, and Strange Quarks

So now our 1970 picture (remember the tau was not discovered until 1975) for quarks and leptons (forget about antiparticles for the moment) is:

u	d		s
ν_e	e^-	ν_μ	μ^-

Charm Quark

Some physicists were struck (as I am sure you are) by the lack of symmetry in this picture. Three quarks and four leptons just did not seem right, and this lack of symmetry led theoretical physicists to predict the existence of a new quark, the **charm quark** (c). It was subsequently discovered in 1974 at the Brookhaven National Laboratory on Long Island and the Stanford Linear Accelerator Center (SLAC) in California. It was not discovered in a free state but bound with the anticharm quark into a meson called J/ψ. From the study of this particle, the properties of the charm quark were determined. It has a mass of 1.5 times the proton mass and a charge of +2/3.

The up and down quarks together with the electron and the electron neutrino form a group called the **first generation** of fundamental particles. The charm and strange quarks together with the muon and muon neutrino form another group called the **second generation** of fundamental particles. But this is not the complete picture on quarks and leptons. As we already know, in 1975 the τ was discovered, and sometime later the corresponding neutrino was inferred. This increased the number of fundamental particles to four quarks and six leptons, and again there was an asymmetric situation:

u	d	c	s		
ν_e	e^-	ν_μ	μ^-	ν_τ	τ^-

Bottom and Top Quarks

So, two more quarks were proposed to complete the picture. They were whimsically named **top** or **truth** (t) and **bottom** or **beauty** (b). The six quark types are commonly referred to as quark flavors. The bottom quark was discovered (also bound with its antiquark into a meson) in the late '70s and was found to have a mass of about 5 times the proton mass and a charge of -1/3. The top quark discovery at Fermilab was announced in March of 1995. One experimental collaboration (CDF) has determined the mass to be about 190 times the mass of the proton and another collaboration (D-Zero) has determined that the mass is about 13% larger than that. So we will say that the mass of the top quark is approximately 200 times the mass of a proton. This discovery completes the model of six quarks and leptons: part of the Standard Model. Together with the force carriers, the photon, the graviton, the gluons (there are actually 8 gluons) and the W's and the Z, we have a model that has been successful in explaining much in the field of particle physics. That does not mean that there are no questions without answers; in fact, there are many; but we will return to some of the questions after discussing accelerators and detectors: the tools of high-energy physics.

u	d	c	s	t	b
ν_e	e^-	ν_μ	μ^-	ν_τ	τ^-

Figure 6-4. The Standard Model.

"Excuse me for interrupting," I said, "but which of you is Up and which of you is Down?" "What? You can't tell? What are you, a muon? I have never been so insulted in my life. I happen to be an Up quark and that is a Down quark and that is another soon-to-be Up quark." "Au contraire, mon frere! You are an Up quark and I am a Down quark and he is staying a Down quark. And further more, I am not just a Down quark, I am a red Down quark and I would advise you to remember that!" "You're red? I can't see any difference among you. Does it make a difference what color you are?" I inquired. "It makes a great deal of difference! I would not be caught dead in the same particle with someone else who was wearing red..." Jeffrey Gordon

Summary

In this chapter:

We were able to get a deeper understanding of:

- The strong force
    - as it rearranges quarks.
    - as it produces quark/antiquark pairs.
    - as quarks and antiquarks annihilate.

- The weak force
    - as it changes a single quark into other another flavor of quark.

We discussed the theoretical motivation for and the experimental verification of:

- The standard model of fundamental particles

    - six quarks.
    - six leptons.

Self-Test 6

For multiple choice, check all that apply.

1. The strong force

   a. can change an up quark into a down quark.
   b. can rearrange quarks.
   c. can make an up/antiup pair from energy.

2. The weak force

   a. cannot rearrange quarks.
   b. can turn a strange quark into an up quark.
   c. can rearrange quarks.

3. The charm quark

   a. was discovered in a free state.
   b. has no electric charge.
   c. was discovered in an accelerator experiment.

4. The bottom quark

   a. has never been seen.
   b. has spin of 1/2.
   c. was discovered in an accelerator experiment.

5. The top quark

   a. is less massive than the proton.
   b. is almost 200 times as massive as the proton.
   c. was discovered in the 60's.

6. Show the quark model, draw a Feynman diagram, and give the force responsible for this process:

$$\pi^+ + n \Rightarrow \Sigma^+ + K^0$$

Answers to Self-Test 6

For multiple choice, check all that apply.

1. The strong force

    a. can change an up quark into a down quark.
    b. can rearrange quarks.   X
    c. can make an up/antiup pair from energy. X

2. The weak force

    a. cannot rearrange quarks.
    b. can turn a strange quark into an up quark. X
    c. can rearrange quarks.   X

3. The charm quark

    a. was discovered in a free state.
    b. has no electric charge.
    c. was discovered in an accelerator experiment.   X

4. The bottom quark

    a. has never been seen.
    b. has spin of 1/2.   X
    c. was discovered in an accelerator experiment.   X

5. The top quark

    a. is less massive than the proton.
    b. is almost 200 times as massive as the proton.   X
    c. was discovered in the 60's.

6. The quark model for the process:

$$\pi^+ + n \implies \Sigma^+ + K^0$$

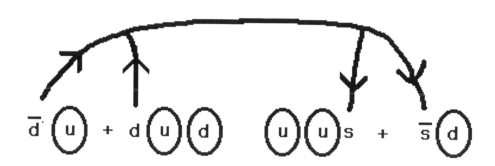

The Feynman diagram:

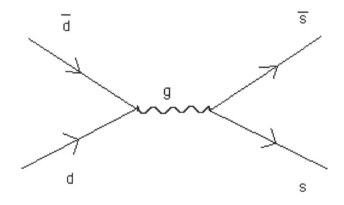

STRONG FORCE

Chapter 7 PARTICLE ACCELERATORS

Particle accelerators are a basic tool of high-energy physics. The majority of the discoveries and theories we have discussed would not have been possible without accelerators and the detectors designed to go with them. In this chapter we will learn about the different types of accelerators:

- linear.
- linear collider.
- fixed target.
- colliders.

Acceleration of Charged Particles

If a charged particle passes through a region where there is an electric field, it will experience a force, as indicated in Figure 7-1. Positively charged particles, like protons, are accelerated in the direction of the electric field: their speed will increase in that direction. Negatively charged particles, like electrons, will be accelerated in the direction opposite to the direction of the electric field, and their speed will increase in that opposite direction. This increase in the speed of a particle results in an increase in the energy (kinetic) of the particle, which is the basic goal of a particle accelerator: to increase the energy of a particle. This increase in energy allows one to study more-interesting particle reactions.

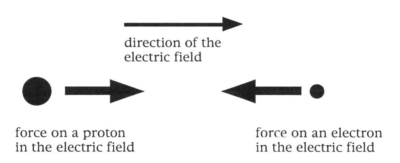

direction of the
electric field

force on a proton
in the electric field

force on an electron
in the electric field

Figure 7-1. Acceleration of charged particles.

"First, if we want really extensive reactions to occur, we need to supply a lot of energy to the reaction system. The best way to do this is for us to achieve high speeds. We can't accelerate to the necessary speeds on our own, so the people place us in electric fields. These make us speed up really fast (those of us with a charged nature anyway)." Christina Lomasney

Particles that are accelerated in high-energy physics are antiprotons, protons, electrons, and positrons ($\overline{p}$, p, e^{+}, e^{-}), for the simple reason that they are stable and have charge. Technically, each one can be used in any of the four types of accelerators, but not all cases are practical.

Linear Accelerators

A linear accelerator (linac) is the simplest type to understand. Particles are injected at one end, are accelerated by electric fields, and come out the other end with a higher energy (as can be seen in Figure 7-2). The longer the linac, the higher the energy of the exiting particle. Many important experiments were done at SLAC (the Stanford Linear Accelerator Center) in California. The SLAC linac is 2 miles long and can accelerate electrons or positrons to 50 GeV (giga electron volts). GeV's are standard units for mass or energy in high-energy physics. To put these units in perspective, 1 GeV is approximately equal to the mass energy of a proton (from Einstein's $E=mc^2$). Electrons emerging from the linac have speeds very close to the speed of light! Protons and antiprotons can also be accelerated in linear accelerators; however, since they are more massive (close to 2,000 times the electron), the acceleration process is more complicated and expensive.

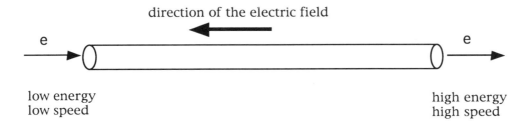

Figure 7-2. Linear accelerator.

Linear Colliders

A new machine began operation at SLAC in 1989 called the Stanford Linear Collider (SLC). The existing linear accelerator was used to accelerate a beam of positrons and a beam of electrons side by side. Magnetic fields were then used to bend the particles along arcs in a newly constructed section to an area where they could collide (see Figure 7-3). In April 1989, the first 100 GeV *collisions* of electrons and positrons were recorded. These collisions were studied, and physicists were able to measure the mass of the Z particle (one of the carriers of the weak force) more accurately. This kind of machine is a single-pass collider; that is, the particle beams have one chance to collide, and then you begin again with new beams.

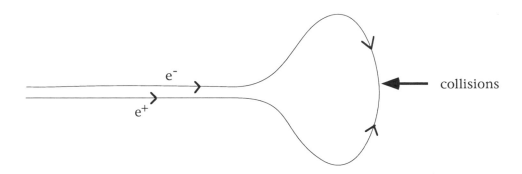

Figure 7-3. Linear collider.

Synchrotrons

A synchrotron accelerates particles using electric fields over and over in a circular path. Magnetic fields are used to bend the particle trajectories and keep them moving in a circle. Proton synchrotrons are more practical to higher energies than electron synchrotrons. When electrons or positrons are accelerated in a synchrotron, they lose so much energy rounding the curves that a large part of the accelerator is taken up with making up this energy loss, rather than accelerating the particles to a higher energy. Beyond an energy of 10 or 20 GeV, a linear accelerator is more economical for accelerating electrons or positrons. When a charged particle is accelerated in a circular path, the particle will radiate (lose) energy. The amount of energy lost depends on how sharp the bend of the circular path is and how massive the particle is. Lighter particles will lose more energy, and this loss will therefore be a much greater problem for electrons (or positrons) than protons (or antiprotons). It will also be less of a problem if the accelerator has a very large circumference. Antiprotons can be accelerated this way, but the difficult part there is making and storing the antiprotons in the first place.

Figure 7-4 shows a simple model of a proton synchrotron. These are also called fixed-target machines, because the beam of moving particles, once extracted, strikes a stationary target. The operation of such an accelerator can be thought of as having three stages:

- Protons are injected into the ring by a linac.

- They circulate until the desired energy is reached.

- They are extracted as a beam (or beams) heading toward targets or detectors.

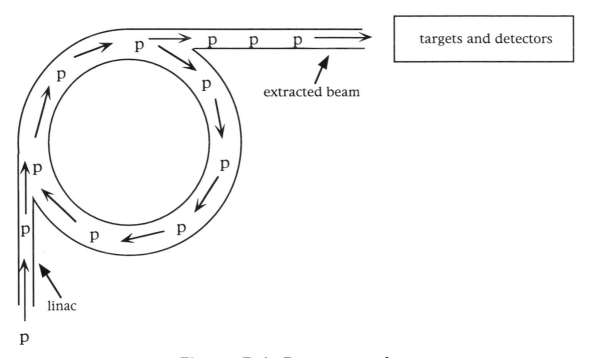

Figure 7-4. Proton synchrotron.

One very useful feature of a fixed-target machine is its ability to make secondary beams, which can be seen schematically in Figure 7-5. When the primary proton beam strikes the stationary target, many different types of particles are produced. The particles shown in Figure 7-6 are just examples, and there are many other possibilities. If a magnet is placed in the path of the particles, the positive ones will bend one way, the negative ones will bend the other way, and the neutral ones will pass straight through. In this manner one can separate beams with different charges. Other techniques can be used to further separate the particles and provide a uniform particle beam. There are proton synchrotrons at Brookhaven National Lab, Fermilab, and CERN, to name a few (see Table 7-2 on Page 88).

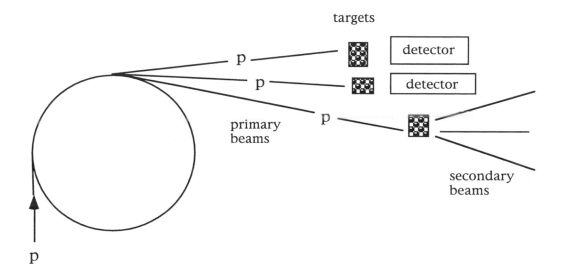

Figure 7-5. Secondary beams.

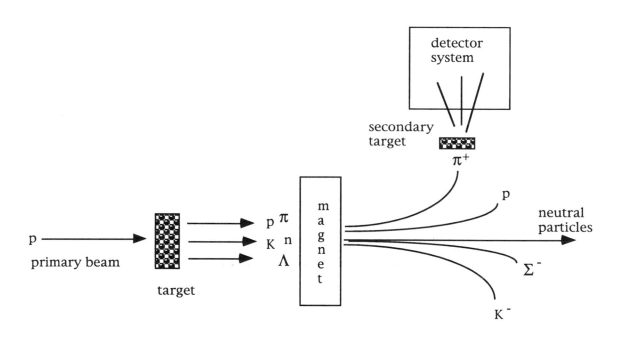

Figure 7-6. Secondary beams (details).

Colliders

The other kind of circular accelerator used in high-energy physics research is the collider. In this type of machine, two beams of particles are accelerated in opposite directions. When the proper energy is achieved, the beams are allowed to cross and collide (see Figure 7-7). Currently the accelerators capable of the highest energy processes are colliders (e^+e^-, pp, e^-p, or p$\bar{p}$). The largest electron-positron collider in the world is called LEP (for Large Electron Positron collider), and it began operating in the summer of 1989 in Geneva, Switzerland. The experiments there have already been successful in placing strong limits on the number of neutrinos and determining the Z mass more precisely. This will probably be the last electron-positron circular collider ever built, because to increase the energy substantially would require a much larger machine, and LEP already has a circumference of 27 kilometers. It may be possible to get higher energy electron-positron collisions with machines like the SLC or two linear accelerators firing particle beams head on. One of the head-on types may be built in the Soviet Union for completion near the end of the decade, with collision energies of at least ten times LEP energies. Much of the interesting physics is expected to come from experiments done at the LHC (Large Hadron Collider at CERN in Geneva), which is scheduled to begin operation in 2004. The table at the end of this chapter gives information on some accelerators of the past, present, and future.

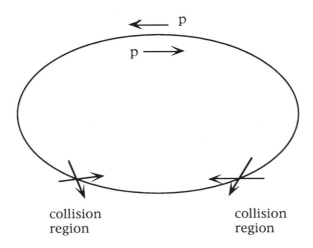

collision   collision
region   region

Figure 7-7.   Collider.

One of the big advantages of a collider is the amount of energy available from the collisions. Imagine a car moving at 30 miles per hour hitting a stationary car and the damage that might be done. Now imagine two cars, each moving at 30 miles per hour, colliding head on. The damage done is most definitely greater. The same principles can be applied to fixed-target and collider accelerators. As can be seen in the table below, you get much more energy for producing new particles in a collider. Furthermore, doubling the beam energy in a collider doubles the available energy, which is not the case for a fixed-target machine. In the table below, energies are quoted in GeV's, where 1 GeV is approximately equal to the mass of a proton.

type of accelerator	energy of proton beam (or beams)	energy available for production of new particles
fixed target	500 GeV	31 GeV
collider	500 GeV	1,000 GeV
fixed target	1,000 GeV	44 GeV
collider	1,000 GeV	2,000 GeV

Table 7-1. Collider vs. fixed target.

"Look out! We're being collected!!" The Helium atoms scattered to avoid the physicists. Lucy the Proton stayed still, excited and expectant. Today was her lucky day. She was pretty sure that she was going to be put into a particle accelerator. The ride to the accelerator wasn't bad. The accelerator turned out to be vast and menacing, a large coil of a snake. There was an ominous detection center of some sort, but Lucy could not quite see it. She cried out as she was separated from her atom and put into a large box. It was there that she met Andrew, another proton. "Are you ready to smash?" he asked. "I heard that they smash us into antiprotons. Imagine that." "Yes I heard," Lucy replied. "You know, most protons would be afraid of antimatter, but for some reason, I'm not. Oh, wow, what if we turn into top quarks?..." Alyssa Babcock

name	type	energy	location	yrs.
Cosmotron	p synchrotron	3 GeV	Brookhaven, NY	1952 - 67
Bevatron	p synchrotron	6.4 GeV	UC Berkeley	1954 - 85
AGS	p synchrotron	28 GeV	Brookhaven, NY	1961 -
SLAC	e^- linac	50 GeV	Stanford, Calif.	1961 -
Fermilab	p synchrotron	400 GeV	Batavia, Illinois	1972 -
CERN S p,$\bar{p}$ S	p,$\bar{p}$ collider	900 GeV	Geneva, Switzerland	1981 - 90
Tevatron	p,$\bar{p}$ collider	2,000 GeV	Batavia, Illinois	1987 -
SLC	e^-,e^+ linear collider	100 GeV	Stanford, Calif.	1989 -
LEP	e^-, e^+ collider	180 GeV	Geneva, Switzerland	1989 -
HERA	e^-, p collider	30 GeV (e^-) 820 GeV (p)	Hamburg, Germany	1992 -
DAΦNE	e^-, e^+ collider	1 GeV	Frascati, Italy	1997 ?
LHC	p, p collider	14,000 GeV	Geneva, Switzerland	2004 ?

Table 7-2. Some past, present, and future accelerators.

Summary

In this chapter:

We learned about four types of accelerators.

- Linear Accelerator

  - Particles are injected at one end and emerge from the other end with a higher energy.
  - Disadvantage: they must be very long to achieve high energies.
  - Example: SLAC linac in California.

- Linear Collider

  - An adaptation on a linear accelerator that allows beams accelerated side by side to collide.
  - Advantage: not as much energy is lost for electrons and positrons as in circular machines.
  - Disadvantage: single chance for collision, therefore beams must be highly focused.
  - Example: SLC in California.

- Fixed Target

  - Electric and magnetic fields are used to accelerate particles (typically protons) around a circular path over and over. When the desired energy is attained, the particles are ejected from the accelerator as a beam.
  - Advantage: secondary beams are possible.
  - Example: Brookhaven Alternating Gradient Synchrotron (AGS) in Long Island, New York.

- Collider

  - Two beams of particles are accelerated in opposite directions and, when the proper energy is achieved, the beams are allowed to cross and collide. Currently the accelerators that are capable of the highest energy processes are colliders.
  - Advantage: high energies available to create new particles.
  - Example: Tevatron at Fermilab in Illinois.

Self-Test 7

For multiple choice, check all that apply.

1. An electric field can be used to accelerate

    a. protons.
    b. neutrons.
    c. positrons.
    d. neutrinos.

2. If a magnetic field bends electrons to the left it will bend

    a. protons to the right.
    b. neutrons to the left.
    c. positrons to the left.
    d. neutrinos either right or left.

3. A linear accelerator

    a. can only accelerate protons.
    b. must be very long to get high energies.
    c. uses magnetic fields to bend particles.

4. A collider accelerator

    a. uses electric and magnetic fields.
    b. can only accelerate protons.
    c. can accelerate antimatter.

5. One advantage of a fixed-target machine is

    a. it uses no electric fields.
    b. it can produce secondary beams.
    c. it can accelerate antimatter.

Answers to Self-Test 7

1. An electric field can be used to accelerate

    a. protons.   X
    b. neutrons.
    c. positrons.   X
    d. neutrinos.

2. If a magnetic field bends electrons to the left it will bend

    a. protons to the right.   X
    b. neutrons to the left.
    c. positrons to the left.
    d. neutrinos either right or left.

3. A linear accelerator

    a. can only accelerate protons.
    b. must be very long to get high energies.   X
    c. uses magnetic fields to bend particles.

4. A collider accelerator

    a. uses electric and magnetic fields.   X
    b. can only accelerate protons.
    c. can accelerate antimatter.   X

5. One advantage of a fixed-target machine is

    a. it uses no electric fields.
    b. it can produce secondary beams.   X
    c. it can accelerate antimatter.

Chapter 8 PARTICLE DETECTORS

To study the particle reactions produced by accelerators, we need a variety of detectors. We don't actually see the particles in the literal sense; however, we can determine many of their properties. In these experiments, some particle properties that physicists might want to determine are:

- velocity
- momentum
- energy

- mass
- charge
- identity

No single detector can do all these things for all particles, so most experiments employ many different kinds of detectors. In this chapter we will learn about some of the detectors used in high-energy physics:

- scintillation counters
- wire or drift chambers
- cloud and bubble chambers
- lead glass
- Cerenkov counters

Scintillation Counters

One of the simplest and oldest types of detectors is a scintillation counter. It is used to give information about the position of charged particles. It can answer the question "Is there anything there at all?" The big advantage of a scintillation counter is that the information is available very quickly, on the order of 10 nanoseconds (a nanosecond is one thousandth of a millionth of a second), but the disadvantage is that the position information is crude. When a charged particle passes through a scintillation counter it stimulates the emission of photons (light) which can then be detected by a photomultiplier tube (see Figure 8-1) The photomultiplier tube puts out an electrical signal signifying the passage of a charged particle *somewhere* in the counter. The size of these counters varies from experiment to experiment, from a few centimeters by a few centimeters to perhaps a meter by a meter.

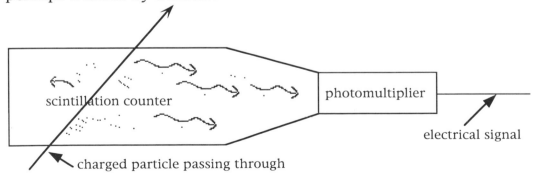

Figure 8-1. Scintillation counter.

Wire, Drift, and Bubble Chambers

To determine the position of a particle more accurately than the scintillation counter does, wire chambers or drift chambers can be used. They also only work for charged particles, and they must be accompanied by fast electronics to record the data. Figure 8-2 shows the basic idea of a wire chamber. Wires are strung millimeters apart in a closed region containing a mixture of gases. When a charged particle passes through the chamber, it ionizes the gas (strips electrons loose). These electrons are then attracted to the wires, which are held at a high voltage (perhaps 2,000 to 3,000 volts). The wire(s) closest to where the particle passed will then have a current signaling the passage of the particle. Therefore, the position of the particle can be known very precisely (within millimeters). Drift chambers are similar to wire chambers, but they also incorporate timing information, making position measurements even more precise.

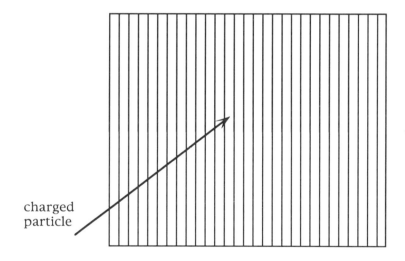

charged
particle

Figure 8-2. Wire chamber.

Position information can allow one to calculate the momentum of a particle. If one sets up a system of these wire chambers with a magnet, one can determine the momentum of the particle by the amount it bends in the magnetic field. The more bend, the lower the momentum, and the direction of the bend gives the sign of the electric charge of the particle. Positive particles bend one way in a magnetic field, and negative particles bend the other.

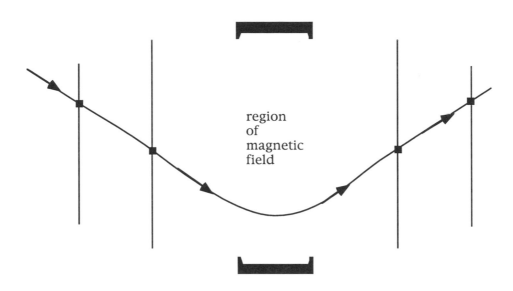

region
of
magnetic
field

Figure 8-3. Tracking in a magnetic field.

Even though they are not used much today, we should learn a little about cloud chambers and bubble chambers, since they did play important roles in early particle discoveries. The pion, positron, and muon discoveries were all made with cloud chambers. A cloud chamber contains gas that is supersaturated (waiting to condense). When a charged particle passes through, droplets form along the path of the particle, leaving a track that can be photographed and studied. Properties of the particle, like momentum, charge, and mass, can be calculated from the photographs. In the 1950s, the cloud chamber was essentially replaced by the bubble chamber, where tracks are produced when a superheated liquid boils along the particle track. Bubble chamber photographs were more precise than cloud chamber photographs.

Neutral particles could be observed in bubble and cloud chambers by observing their decay products. Figure 8-4 shows an example of tracks that may be seen in a bubble chamber. You can see that the two tracks on the upper left seem to come out of nowhere. Usually these kinds of tracks can be attributed to a neutral particle coming from the other vertex. Careful measurements of these tracks reveal that they are from the decay of a cascade into a negative pion and a lambda that subsequently decays into a proton and a negative pion. The particle assignments can be seen in Figure 8-5.

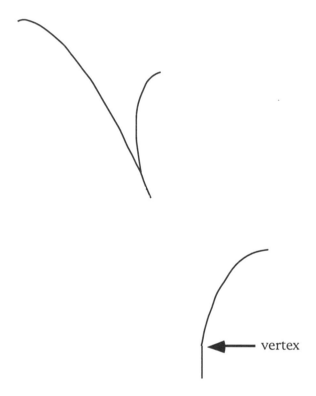

Figure 8-4. Bubble chamber tracks.

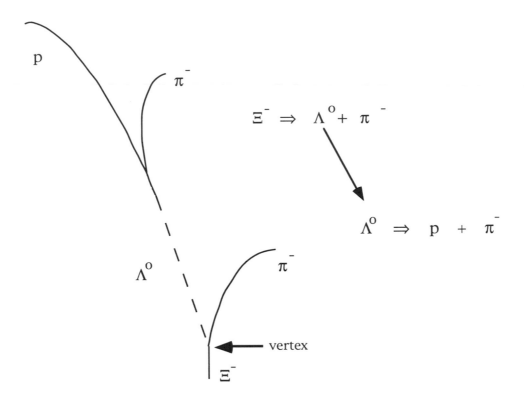

$$\Xi^- \Rightarrow \Lambda^o + \pi^-$$

$$\Lambda^o \Rightarrow p + \pi^-$$

Figure 8-5. Bubble chamber tracks revealed.

"<u>Question</u>: You've often been characterized as a secretive particle. The fact that you personally have never appeared in a bubble chamber photograph, leaving only the V vertex shape, has only added fuel to the fire. For posterity, what makes a neutral kaon one of the strangest of the strange particles tick?

<u>Answer</u>: About all the secrecy, there's a theory in show business that says overexposure is as bad as underexposure, so I am a little careful about that. As for never doing bubble chamber pictures, I'm sorry, but I'm a private particle. If I wanted to be a publicity hound, I would have electrical charge. If that were the case, O.K., I'd show up in the pictures. But I'm not charged, I am neutral. I don't think that's unfair of me, given that both my parents and my children do those bubble chamber shots all the time. How many times does a particle need to be seen before you know that it is there? I mean, my parents are in the pictures and my kids, the pion twins, make the V vertex. That's like my calling card, the mysterious V vertex..."

Michael Sheresky

Lead-Glass Detector

Photons are neutral particles that are abundant in high-energy physics. One can sometimes infer the existence of a photon by observing production of an electron/positron pair, but that is not always the case. Determining the energy of photons is often of importance in experiments, and this can be accomplished using a detector called lead-glass. When photons (or electrons) encounter a special type of glass called lead-glass, they deposit their energy in the glass. A photomultiplier tube attached to the glass produces an electrical signal proportional to the energy of the photon or electron. The more energy deposited by the particle, the greater the electrical signal from the photomultiplier tube. The figure below shows one piece of a lead-glass detector system.

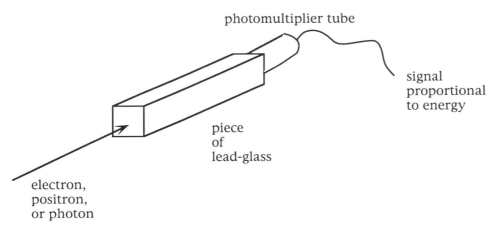

Figure 8-6. Lead-glass detector.

Electrons can be distinguished from photons by the presence of an associated charged particle track in another part of the detector system. Detectors that measure energy of particles are called **calorimeters**. Lead glass is a calorimeter for photons and electrons (therefore called an electromagnetic calorimeter). There are calorimeters specifically designed to measure energies of different types of particles. Hadron calorimeters are used to determine energies of hadrons (like pions and protons). Figure 8-7 shows a lead-glass array (consisting of 256 pieces stacked in a 16-by-16 square array), looking from the direction of the incoming particles. Each square represents one piece of lead glass. The photomultiplier tubes are on the back sides of the glass. The example shown indicates that three particles entered the glass array: The one with the smallest energy is near the center, the next most energetic is near the top left corner, and the most energetic is near the bottom left.

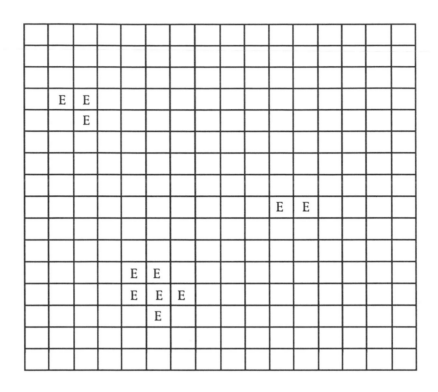

Figure 8-7. Lead-glass array.

"Well, I'm glad to see you made the adjustment," said the voice which no longer sounded so small or peculiar. "Allow me to introduce myself, I am Ernest, the Electron Neutrino. I am perhaps the smallest of all the subatomic particles you could meet. It's because I am so tiny (and I don't interact much) that I could go through the many layers of steel shielding that surround the particle detectors. Not many particles can go anywhere that they please you know. And I am not burdened by a charge, so I can't be pushed around by those silly magnets. I get to most anything I want. Mostly I just go to particle decays. There's not much else to do once you're on the subatomic level..." Jeffrey Gordon

Cerenkov Counters and Particle Identity

Particle identities can be determined from various kinds of information. One way to determine the type of particle is to determine its mass. A simultaneous measurement of the momentum and energy of a particle will enable the calculation of its mass. Another way to determine the type of particle is to make use of a Cerenkov counter to measure its velocity. This kind of counter works because particles moving through a material (called the **radiator**) will emit light if their speed in the radiator is greater than the speed of light in the radiator. This will be possible because the speed of light in a material is not as great as the speed of light in a vacuum (the maximum possible speed attainable). The key point here is that the light emitted by the particle is emitted at an angle that depends on the speed of the particle. The higher the speed, the larger the angle of radiation (as measured from the original particle direction). So, if you have a beam of particles whose momentum is known, the Cerenkov counter will allow you to identify particles of different mass. Figure 8-8 shows how one might design a Cerenkov counter to give a signal for kaons, but not pions or protons. If you wanted to identify protons, you would place the mirror in a different place. The pion mass is smaller than the kaon mass, so its speed is greater. Therefore the angle of the emitted light is greater than the angle of the light emitted by the kaon, and it misses the mirror. The proton, which is more massive than the kaon, has a smaller velocity and, hence, a smaller angle for emitted light, which also misses the mirror. Figure 8-9 shows a close-up of the radiation emitted.

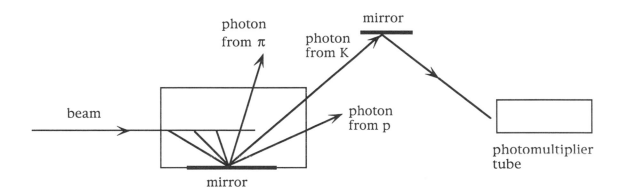

Figure 8-8. Cerenkov counter.

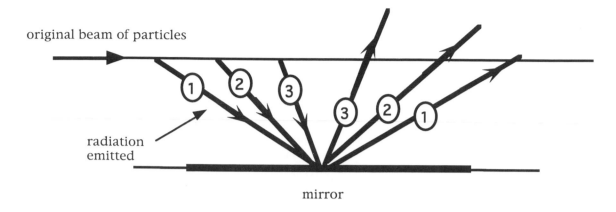

original beam of particles

radiation
emitted

mirror

1 is a photon emitted from the proton
2 is a photon emitted from the kaon
3 is a photon emitted from the pion

Figure 8-9. Cerenkov counter details.

Another way to identify particles is by their unique properties. For example, muons can pass through a lot of material (like steel), but electrons and pions cannot. So, if you can identify a charged particle before it passes through a lot of steel and again after it passes through, you can be fairly sure that it is a muon. Figure 8-10 illustrates how one might distinguish muons from pions and electrons.

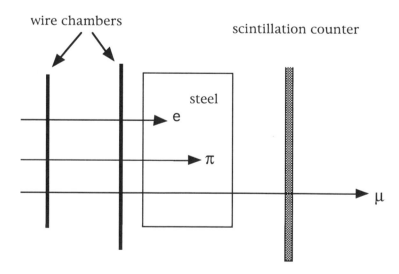

wire chambers

scintillation counter

steel

e

π

μ

Figure 8-10. Particle identification.

Summary

In this chapter:

We learned about some of the detectors employed in high-energy physics and their primary uses:

- Scintillation counters give quick, crude information about particle position.

- Wire and drift chambers give more accurate information about particle positions. When combined with a magnet, they can determine the momentum of charged particles.

- Cloud and bubble chambers are not used much today, but were crucial to the discoveries of particles like the positron, muon, and pion. They allow one to see charged particle tracks and determine properties like momentum charge and lifetime.

- Lead glass is used to measure photon and electron (or positron) energies.

- Cerenkov counters are used to identify particles by measuring their velocity.

Self-Test 8

For multiple choice, check all that apply.

1. Scintillation counters can be used to

      a. measure momentum.
      b. measure position crudely.
      c. measure position precisely.

2. Lead glass can be used to

      a. measure photon energies.
      b. measure positron energies.
      c. measure proton energies.

3. Wire chambers can

      a. be used to measure charged particle position.
      b. be used to help determine the momentum of a particle.
      c. directly measure electron energy.

4. Particle velocity can be determined from

      a. a Cerenkov counter.
      b. a single drift chamber.
      c. lead glass.

5. Draw a bubble chamber picture representing the decay

$$K^+ \Rightarrow \pi^+ \pi^0$$

with the π^0 decaying into two photons, which both pair-produce an electron and a positron.

6. Design a detector system to observe the decay

$$K^0 \Rightarrow \mu^+ + e^-$$

Hints:

You need to determine the momentum of each particle.
You need to determine if the muon *is* a muon.
You need to measure the energy of the electron.

You will need to use:

    4 wire chambers
    1 magnet
    lead glass
    scintillation counters
    steel

Answers to Self-Test 8

1. Scintillation counters can be used to

    a. measure momentum.
    b. measure position crudely.   X
    c. measure position precisely.

2. Lead glass can be used to

    a. measure photon energies.   X
    b. measure positron energies.   X
    c. measure proton energies.

3. Wire chambers can

    a. be used to measure charged particle position.   X
    b. be used to help determine the momentum of a particle.   X
    c. directly measure electron energy.

4. Particle velocity can be determined from

    a. a Cerenkov counter.   X
    b. a single drift chamber.
    c. lead glass

5. Draw a bubble chamber picture representing the decay

$$K^+ \Rightarrow \pi^+ \pi^0$$

with the π^0 decaying into two photons, which both pair-produce an electron and a positron. Notice how the positive particles bend one way and the negative particles bend the other way. Your picture will probably not look exactly like this, but the features should be similar.

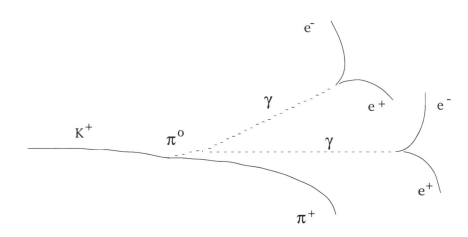

6. Design a detector system to observe the decay

$$K^0 \Rightarrow \mu^+ + e^-$$

To determine the momentum of each particle, the wire chambers are placed in front of and behind a magnet. The positions measured in the chambers allow calculation of the bend, which gives the momentum. You will need to see one particle bending to the left and one to the right.

To determine if the muon *is* a muon, one can use a combination of scintillation counters and steel. Particles like pions and electrons, which could be confused with muons, will not make it through the steel.

To measure the energy of the electron, one needs an array of lead glass.

To further differentiate pions, electrons, and muons, a Cerenkov counter can be inserted as well.

Note that this is not the only possible design. You may have come up with something that looks a little different but accomplishes the same goal. See the drawing on the next page for this design (which was a real experiment at Brookhaven).

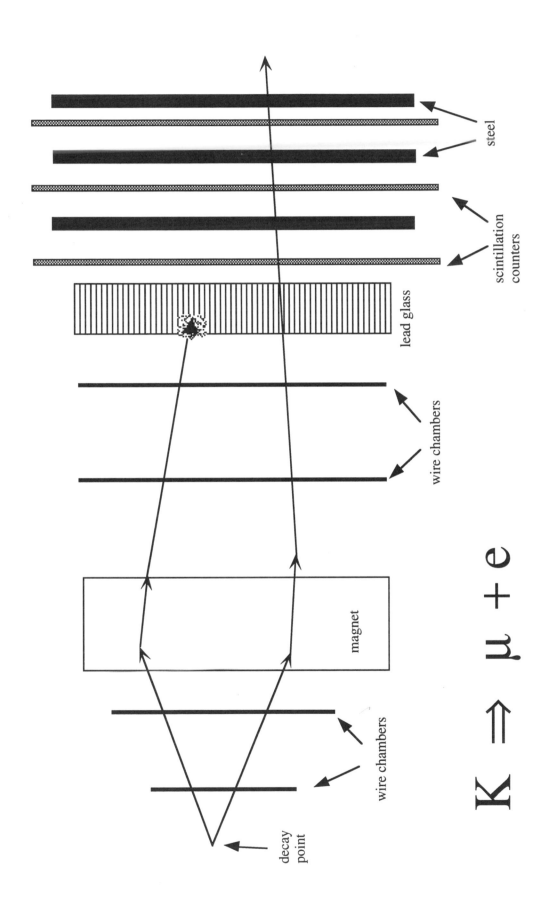

steel

scintillation counters

lead glass

wire chambers

magnet

wire chambers

decay point

$$K \Rightarrow \mu + e$$

Chapter 9 Open Questions

We have seen almost 100 years of theories and experiments leading up to the present standard model of the fundamental entities that make up the world around us. The most recent testament to its accuracy and success is the discovery in 1995 of the top quark. Even at the writing of this book there is some data from the CDF collaboration that may be interpreted to indicate some sort of substructure to quarks. By the time it gets to print this may (or may not) be shown to be the case. There is still much more work to be done in getting a more accurate value for the mass of the top quark and studying how it decays. The Standard Model will continue to be tested in many places around the world even as you read this. The Standard Model may be replaced in the near (or distant) future, but more likely it will end up being modified to include new discoveries and/or new theories. Experiments are going on all over the world and theorists are hard at work. Many high-energy physicists are planning the experiments to be done at the LHC (large hadron collider). There are many questions still to be answered and many more to come that we can not imagine today. Just a few of the questions foremost on the minds of high-energy physicists are:

- What is the origin of mass? Some leptons seem to have little or no mass, yet all quarks do. Why is this the case? And why do the quarks have the masses that they do? Are neutrinos massless?

quarks		leptons	
u	≈ .005 GeV	e	≈ .5 Mev
d	≈ .01 GeV	μ	≈ 100 MeV
s	≈ .2 GeV	τ	≈ 1800 MeV
c	≈ 1.5 GeV	ν_e	< 7 eV
b	≈ 4.7 GeV	ν_μ	< .3 MeV
t	≈ 180 GeV	ν_τ	< 30 MeV

Remember, an MeV is one-thousandth of a GeV. And an eV is one-millionth of an MeV.

- How many generations of quarks and leptons are there?

The latest experiments from LEP and SLC say only three, but why?.....

u	d	e^-	ν_e
c	s	μ^-	ν_μ
t	b	τ^-	ν_τ

Completing the three generations with the discovery of the top quark was key and it was awaited for almost two decades. What will we see next?

- And what about observing free quarks?

  The present theories support permanent confinement inside of hadrons, that is we may never see a free quark.

- Are quarks and leptons composed of parts?

  We have seen the structure of matter probed further and further.
  From the atom
          to the nucleus
              to the proton
                 to the quark
                    to the.....

the atom:
once thought to be
indivisible, now known
to have parts, the nucleus
and electrons

the nucleus:
once though to be
a solid entity, now known
to have parts, protons and neutrons

the proton:
once thought to be
fundamental, now known
to have parts, quarks

the quark:
now thought to be
fundamental ???

Figure 9-1. Parts within parts.

Appendix

My Course Spring 1995

Required Reading: A Tour of the Subatomic Zoo - in the college store

Recommended Reading: Packets from Fermilab - I will hand out to you

Required Work:

Five one-page essay questions due as indicated on the course syllabus.

Three graded activities (2 for homework and one in-class) due as shown on the syllabus.

Final Paper: Floating due date May 4 -May 9. Must be in by 3:00 on May 9. This is a firm deadline. <u>No papers will be accepted after that time without a written excuse from the appropriate class dean or class advisor and you will receive an F in the course.</u>

Paper choices (all must be typed):

1. Fictional story with subatomic particles as characters. (3-4 pages)
2. Poem about subatomic particles. (1-2 pages)
3. The SSC - its conception and its ultimate demise. (3-4 pages)
4. The search for and the ultimate discovery of the top quark. (3-4 pages)
5. Uses of accelerators in medicine. (3-4 pages)

Your final grade will be determined by the following:

- up to 50 points for the 5 one-page essays (10 for each, 0 for any not handed in on time)

- up to 25 points for the 3 graded activities (from CPEP packet)

- up to 25 points for the final paper

All work must be handed in on time. You can not pass the course without handing in the final paper, even if your total from the rest of the graded work is numerically above the passing cutoff.

Hope that you enjoy the course,
Professor Cindy Schwarz

Syllabus for Physics 168

Date	Topics Covered	Work due	Reading
3/30	introduction, history of matter views, atoms, Rutherford, nucleus, uncertainty principle		chapter 1
4/4	Video - Race for the Top	10 Questions about the video collected	
4/6	protons, electrons, spin, neutrons, neutrinos, conservation laws		chapter 1
4/11	forces, mediators, Feynman diagrams, antineutrinos, pions, muons, different neutrinos, particle classification	Essay 1 due Receive Activity Two	chapter 2 and 3
4/13	antimatter, leptons, hadrons	Activity Two due	chapter 3
4/18	summary, all leptons, p, n, pions, kaons, lambda, sigma, cascade, charge, lepton #, mass, strangeness, baryon #	Essay 2 due Receive Activity Five	chapter 4
4/20	Eightfold way, quarks	paper abstract due	chapter 5
4/25	particles/antiparticles, quarks and spin, strong vs. weak force, beta decay, quark masses, standard model	Essay 3 due	chapter 6
4/27	QCD, color, gluons, jets	Activity Five due	
5/2	standard model, charm, bottom, tau's, top	Essay 4 due	chapter 6
5/4	accelerators Video - Powers of Ten	Essay 5 due	chapter 7
5/9	detectors Lingering Questions	Final paper due Activity Seven - in class	chapter 8 and 9

Suggestions for Further Reading

<u>books</u>

Close, F. *The Cosmic Onion*, American Institute of Physics, New York, 1986.
A level or two up in mathematical complexity and somewhat more inclusive.

Close, F., M. Marten and C. Sutton. *The Particle Explosion*, Oxford University Press, Oxford, 1987.
More inclusive, much longer and with fantastic color pictures.

Dodd, J. E. *The Ideas of Particle Physics: An Introduction for Scientists*, Cambridge University Press, Cambridge, 1984.
For those with a scientific background who can handle a high level of mathematical sophistication.

Kane, Gordon. *The Particle Garden: Our Universe as Understood by Particle Physicists*, Addison Wesley, (1995), 224 p.

<u>articles</u>

Boslough, J. "Worlds within the Atom," *National Geographic*, Vol. 167, No. 5, May 1985.

Rubbia, C. and M. Jacob. "The Z^o," *American Scientist*, Vol. 78, November 1990.

Lemonick, M. , "The Ultimate Quest," *Time*, April 16, 1990.

<u>other</u>

Contemporary Physics Education Project (CPEP) materials include
    teacher's packet with activities
    Color particles software
    world-wide web site (http://pdg.lbl.gov/cpep/adventure.html)

Fermilab web site (http://fnnews.fnal.gov)

Since the web is expanding constantly, the above sites may move. For updated lists of web sites on relevant material visit my home page where you will find links (http://noether.vassar.edu/~schwarz)

Glossary

annihilation: When a matter/antimatter pair of particles meets and turns into energy.

antiparticle: An antiparticle has the opposite properties of its particle (like charge and strangeness, for example) when there is an opposite, and the same value of properties that do not have possible opposites (like mass).

baryon: A hadron with half-integer spin. All baryons are composed of three quarks.

beta decay: When a neutron decays into a proton, an electron, and an antineutrino. The underlying process is a down quark changing into an up quark, an electron, and an antineutrino. The weak interaction is responsible for beta decay.

cascade: A strange baryon.

collider: An accelerator in which two beams of particles circulate in opposite directions and collide head-on.

conservation: A conserved quantity is one that has the same value before and after a process. Examples of quantities that are conserved in various types of interactions are baryon number, charge, energy, lepton number, momentum, and strangeness.

decay: The process of one particle becoming two or more particles.

detector: Any device that can sense the presence of a particle and give information about one or more of its properties.

electromagnetic force: Force that acts between all charged objects. It can be attractive or repulsive.

electron: A fundamental particle with negative electric charge. It is one of the three constituents of the atom.

force: That which governs the interaction between particles. There are four known fundamental forces: electromagnetic, gravitational, strong, and weak.

force carrier: The particle exchanged during an interaction.

generation: Two leptons and two quarks together form a generation. The electron, its neutrino, and the up and down quarks form the first generation. The muon, its neutrino, and the charm and strange quarks form the second generation. The tau, its neutrino, and the top and bottom quarks form the third generation.

gluon: The mediator of the strong force.

graviton: The mediator of the gravitational force.

gravitational force: The force that acts between all objects with mass. It is always attractive.

hadron: Any particle that experiences the strong force.

kaon: A strange meson with about half the proton mass.

kinetic energy: Energy of motion. Only objects that are moving have kinetic energy.

lambda: A strange baryon.

lepton: Considered to be a fundamental particle. There are six known leptons: the electron, the muon, the tau, and three neutrinos. They do not experience the strong force.

lifetime: The time that an unstable particle (or atom or nucleus) lives before decaying into other particles (or atoms or nuclei).

linear accelerator: A machine that accelerates charged particles in a straight path.

meson: A hadron with integer spin.

muon: A fundamental lepton with a mass of about 200 times the electron mass.

neutrino: A fundamental lepton that has no electric charge and little or no mass. There are three kinds of neutrinos: electron neutrino, muon neutrino, and tau neutrino.

neutron: A constituent of the nucleus. It has no electric charge and is made of three quarks: two downs and one up.

nucleon: The collective name for protons and neutrons.

nucleus: The densest part of an atom, it contains protons and neutrons.

pair production: The opposite of annihilation, when electromagnetic energy becomes a pair of particles.

photon: The carrier of the electromagnetic force.

pion: A meson with a mass of 1/7 the proton.

positron: The antiparticle of the electron.

proton: A constituent of the nucleus. It has positive electric charge and is made of three quarks: two ups and one down.

quark: Considered a fundamental particle according to the standard model. There are six flavors of quarks: up, down, strange, charm, bottom, and top.

radiation: That which is emitted from an atom, nucleus, or particle. Alpha radiation is the emission of a helium nucleus, beta radiation is the emission of an electron (and antineutrino), and gamma radiation is high-energy photon emission.

reaction: Two particles interacting to produce one or more particles.

scintillator: A material that emits light when struck by a charged particle. Scintillation counters are used to detect charged particles.

sigma: A strange baryon slightly more massive than the proton.

spin: An intrinsic property that a particle may possess. Leptons and quarks, the fundamental particles, have 1/2 unit of spin.

Standard Model: A model of six quarks and six leptons as fundamental entities.

strong force: The force that acts between all nucleons. It is attractive for all combinations of protons and neutrons. Quarks feel the strong force, but leptons do not.

tau: A fundamental lepton with a mass of about 3,600 times the electron mass.

W mediators: The charged carriers of the weak force.

weak force: The force that can change one quark type into another. It is the only force affecting neutrinos.

Z mediator: The neutral carrier of the weak force.

Index

P

pair production, 31, 114
particle lifetime, 31
particle momentum, 9
Pauli, 10
photomultiplier, 92, 97
photon, 20, 21, 23, 75, 92, 114
    detection, 97
pion, 31, 32, 34, 35, 62, 114
    decay, 33
    properties, 31
positive muon, 34
positive tau, 34
positron, 29, 34, 35, 47, 115
Powell, 31
proton, 3, 4, 5, 13, 34, 62, 115
    spin, 5

Q

quark, 20, 57, 58, 115
    bottom, 75
    charm, 74
    colors, 63
    confinement, 64
    down, 57, 63
    experimental evidence of, 64
    flavors, 57, 75
    generations, 74
    masses, 108
    strange, 57, 63
    top, 75
    up, 57, 63

R

radiation, 115
    alpha, 7
    beta, 7
    gamma, 7
reaction, 115
Rutherford, 1

S

scintillation counter, 92, 101
scintillator, 1, 115
secondary beams, 84, 85
sigma, 45, 63, 115
    properties, 45
spin, 5, 59, 63, 115
standard model, 108, 115
strange particles, 44
strange quark, 57, 63
strangeness, 49, 50

T

tau, 34, 115
tau antineutrino, 34
tau neutrino, 34
Tevatron, 89
Thomson, 2
top quark, 75
    discovery of, 75

U

up quark, 57, 63

V

velocity
    measurement of, 99

W

W particle, 20, 24, 115
wire chamber, 93

Y

Yukawa, 31

Z

Z particle, 20, 24, 83
Zweig, 57